I0468624

Vaccine Voodoo

Denouements & Discussions - Books 2 & 3

Book 2
Further Voodoo Science

Note: These additions came after I first said "I've finished".

"The Immunobiology Section", at the end of Book 1, was meant as the scientific clincher. It was resolution to doubts and inconsistencies I had held originally when we decided there was no reason to subject our kids to any vaccines. But the Nobel Prize did not next day arrive on my doorstep. There was no sudden mass acceptance of the veracity of my observations. In fact the World did not change. As I mulled further and then mulled again my thinking was attempting to quantify the nature of the establishment and to devise mechanisms to bring home realities. So I carried on writing................

Voodoo Science Extra musings 2014-2015

Contents

1 - Bad Science? Very Bad Science!

February 7, 2014

Thinking once more of that journalist/doctor who supports MMR, hates homeopathy, name of Goldacre. His column "Bad Science" was weekly in The Guardian, picking on herbal remedies, alternative health gurus, homeopaths and anyone who questions vaccination. Ever. At all. In any form.

Anyway, seeing one such article, I kind of went off on an "if" moment. Just think there'd been no vaccines, no Jenner, but we'd got here by a different, vaccine free, route.

Then "one day, rumour has it, there is a new way of tackling childhood infections"............

"Hey. This has been such a roller coaster. I've been able to bring to task so many charlatans and rogues in my column and all my resultant books and stage performances. I have brought to your attention these quaint little individuals who, in the name of providing cures for legitimate illnesses and many imagined ones, too, just work to make me despair. I have depicted larger, long term scams as well and found that even Her Majesty the Queen is not immune to being conned, with her whole family still, albeit covertly, using the absurd practice of homeopathy to cure sufferers by giving them little pills containing utterly nothing. You know – ziltch. Just pellets of chalk.

"Anyway, this week I think that I have come up with a little guy who is doing something so outrageous, so hurtful, harmful and potentially deadly that I feel I should not use any of my usual banter. He is so dangerous, so potentially deadly that he has to be stopped in his tracks. Simply, he is poisoning his clientele by adding uncontrolled cultures of bacteria, fungi and other microorganisms directly into gashes he carves into the recipients flesh with unsterile knives. Incredibly he collects these micro-organisms in the form of puss drawn from boils formed on cattle suffering from the Pox.

"Unbelievably, this total madman plies this trade on the basis that he "Did it to a farmer's son and he is still alive." Oh lucky child, indeed, for such treatment would have packed away many an adult. Apparently he says that this voodoo treatment, this sorcery will prevent the recipient from developing Smallpox but

I fail to see ANY logic behind this. When have humans ever developed cowpox? He'll be herding us in fields soon and leading our wives to the milking parlour of an evening....

"Already, though, there are reports of deaths and disabilities arising from the use of his method, for he has convinced several other practitioners to also follow his ideas. This has to stop now as we have all the capabilities to treat natural illnesses in children and in the rare cases that adults fall ill as well. The detailed science of interactive medical ecology, sometimes, albeit inaccurately, termed "Immunobiology", has detailed the incredibly complex set of interactions between humans and their symbiant bacterial communities. It has shown how improved nutrition and housing and stable social structures so boosted human physiologies that they no longer succumbed to invasive imbalances arising when normal symbiants or, indeed, invasive species could trigger illness. The elegant details of this system of homeostasis, developed over so many years of evolution seems second only to neuro-science in its subtleties and wide capabilities.

"But now, out of the blue, this crazed, I'm going to call it as it is, this crazed child-molester, a man named Jenner, thinks he can do better than that. He should at once be hauled up in front of a court, tried, convicted and then locked away for the rest of his days.

"We cannot allow his idea to gain ANY traction."

2 – Sometimes you just gotta laugh

May 14th, 2014

On Al Arabiya News Sunday, 11 May 2014

The headline was **"Experimental vaccine for MERS developed"**. It went on:

"The experimental vaccine is based on a platform for a candidate that is said to protect against SARS." [*They are not gonna commit here...*]

"A U.S. biotechnology firm and university researchers have reportedly created an experimental vaccine against the Middle East Respiratory Syndrome (MERS), according to media reports." [*.......It is suggested that the virus has mutated under the influence of very high concentrations of affluence in the areas of the Middle East where it has arisen.(?)*]

"Maryland University and Gaithersburg biotech Novavax announced that the vaccine for combating coronavirus has succeeded in stopping infection in laboratory studies, the Saudi Gazette reported on Sunday." [*Wall Street reported sharp increases in Gaithersberg stock price. Surely!*]

"Saudi Arabia has been the country most affected by the outbreak of the MERS virus." [*Due to the virus' requirement to grow in areas of extreme affluence, of course.*] "The number of fatalities on Saturday reached 139 since MERS first appeared in the kingdom in 2012, the Saudi health ministry announced." [*Although sub-human lackies can also develop the virus as long as they live close to the sources of affluence. They are rarely offered care as they are, of course, totally expendable.*]

"According to the Gazette, the experimental vaccine is based on a platform for a candidate that is said to protect against Severe Acute Respiratory Syndrome, or SARS."

Both MERS and SARS are coronaviruses. [*Whether there is a connection to the consumption of soft effervescent beverages is not yet clear.*]

"You're blocking the attachment of the antibody to the human cell," Dr. Gregory Glenn, the senior vice president of research and development at Novavax was quoted as saying by the Gazette. [*Again the meaning of this statement was unclear. It could be that he was referring to the detention of aforementioned lackies – though whether he'd refer to them as "human" is doubtful.*]

Thus he may have been indicating his total lack of understanding of any immunological theory.]

"When you have an immune response to the virus, the virus is destroyed," Glenn added. *[Which naive simplicity appears to emphasise the latter interpretation!]*

Gale Smith, vice president of vaccine development at Novavax, said in a statement carried by Washington Business Journal that the biotechnology firm will "continue to evaluate" the vaccine. *["We have no idea what it does and need to check out its mortality rate."]*

"Novavax will continue to evaluate this technology to produce highly immunogenic nanoparticles for coronavirus, influenza, and other human disease pathogens with the potential for pandemic and sustained human to human transmission," Smith said. *[Here, at least, he is very upfront about the company's work! To paraphrase his words: "We shall continue to work to create new diseases in every devious way we can imagine."]*

U.S. patient makes recovery **[My, we can't have that now, can we. What were those nurses thinking of…..]**

The news comes after the first American diagnosed with the mysterious virus *[How romantic!]* was released from an Indiana hospital and is considered fully recovered, the hospital said Friday, according to The Associated Press. Community Hospital chief medical information officer Dr. Alan Kumar said the patient now tests negative for MERS and "poses no threat to the community." *["See, we've exorcised this devil".]*

The U.S. Centers for Disease Control and Prevention said the patient is an American man. He flew from Saudi Arabia to Chicago on April 24 and took a bus to Indiana. He sought treatment last Monday and was diagnosed with MERS. **[What was that?**
Oh yes, "Mysterious and Evil but Romantic Syndrome………]

3 – Nadine's Cri de Coeur
November 22nd, 2014

I'll just add this copied out as she wrote it. Verbatim:

"I wanted to reach out.

I do not vaccinate.

I try not to judge those that do.

I try to inform but not preach and always be respectful.

My friend had a baby last September.

He was crawling, standing, beginning to walk, saying the typical words, eye contact, normal moods and normal sleep patterns.

He had his vaccines at his one year checkup.

He has stopped crawling, will not stand, will not make eye contact, will not say ANY of the words that he was saying previously.

Has odd fits of "rage" and hits and punches things.

Makes sudden movements and thrashes out of nowhere.

He has stopped eating…it's almost like he has forgotten how to, and will only drink milk from a cup.

He will not sleep but for short periods of time and it is not in his crib as it previously was.

My friend is devastated, as much as I am screaming "I told you so" on the inside,

I would never say that to her.

Her pediatrician is saying that sometimes babies regress....

BULLSHIT

He just happened to regress within 24 hours of getting his vaccines?

Where do I tell her to go?

What do I tell her to do?

I have never vaccinated my boys.

I never intend to.

THIS IS AN EPIDEMIC.

Why are our screams being silenced?

This is a child.

Now after vaccines he will never be the same.

It's a sin, it's a shame, it's COMPLETELY preventable.

Sad."

[Nadine, posted to VINE, November 2014]

4 - When did high tech science get grafted into the Vaccine Voodoo? December 4th, 2014

OK, here's another conceptual difficulty with generic Jennerism (as hinted on page 7). When did a country quack's diabolic corporeal carving and inpasting of putrefied puss stop being the clear and imminent danger to public health and morph into what is now a hi-tec generated impenetrable fog of mystic, deeply scientific writings, disguising the fact that they are still carrying out exactly the same process?

No, what I really mean is that if Dr Jenner were to appear today and even suggest any of the practices he carried out he would be brought before the Star Chamber of the General Medical Corporation and be subject to the very worst of their incantations before being burned at their professional stake. His career would undoubtedly be at an end, his reputation in tatters and he would lose all his friends and ex-colleagues. Probably he'd head off to America but, hey, that's another story!

Somehow, in the interim two hundred years, instead of its being lambasted as a crazy intrusion on anyone's good health and consigned to the deepest dungeons of "terrible things we used to do", like sending young kids to work in cotton mills or soldiers to "fight" in the trenches or using toxic heavy metals like mercury in household potions, instead of that we have elevated this insanity to Global Saviour status and are ever extending its import and application. Based on no plausible evidence and utterly no open investigation of its oh so common collateral damage, vaccination is a medic's favourite tool.

And, of course, "Big Pharm" loves it to pieces as a multi billion dollar mint.

Always the process has killed willy nilly. Death tolls from Jenner's actions far outweighed any theoretical "saving" of folk from the pox. Some would say Jenner's wholly unsterile and clearly dangerous practices were the pox – certainly horrendous reactions would result from any such "medicine" today.

Any member of the GMC want to stand in line so's I can cut a slit in his arm – nice and deep – and then rub some septic, infected pus into the slit? Then I can bind it in a piece of cloth and you can go away and say you won't get ill from the pox. Might just get septicaemia, mind, but, hey, who cares about such a trivial detail. Yeah, I suppose it might kill you but, look, you won't get the pox! Come on now, roll up – you know, well you say, it makes sense. Look, you have even named an institute after this charlatan – in Oxford is now the multimillion pound construct named "The Jenner Institute" to develop more of this voodoo.

So my question is how did this ever get to be "scientific"? There's two aspects to the use of science today. One is simply the ability to apply methodologies to systems. This is process and not discovery. It is closed system operation, with defined constraints or borders, even. Like exploring a small island, maybe and so saying you understand the world. Whatever, it is not open use of scientific method where you are looking always for receipt of new information and ready to reassemble your understanding of the systems you are examining.

Clearly vaccinology – aka "Jennerism" – is stranded on a very small island. They worship the holy Jenner for He saw the future economy of health. Oops, no I meant to say "for he showed us how to fight the menace of the pox". As I suggest above, his methods were such that he could not have achieved such goals and would, equally clearly, have dispatched many, many people to an early grave.

But such was the power of central governance – even then – and such was the stubborn need of the GMC – even then – not to look a complete charlatancy that the concept became ingrained into medical practice. A glance through "Victorian Pharmacy" by Jean Eastoe will give you an understanding of so many other fallacies of the era, of so many lotions and potions long since banned from use that you wonder anyone survived the period at all.

Sadly, though, Jennerism was disguised. So, in about 1860, the first syringe was developed, so obviating the practice of making a gash in the arm – now it was a needle thrust into your skin. "Sterile?" Why of course not, for there was no understanding of microbial infection at the time. People still died in cohorts – whyever would they not?

Into the twentieth century and it went on much the same. Sterilisation arrived – Pasteur, yeah, but perhaps more important was the guy who insisted doctors wash their hands to avoid maternal death from "puerperal fever" straight after childbirth. (A doctor named Semelweiss.) Otherwise the only change was the early perceived desire to extend vaccination beyond simply smallpox. Pasteur tried rabies and a TB jab was formulated. It is strongly suggested, from a host of archive materials, which I am very inclined to agree with, that the bulk of the mega "flu" "epidemic" post World War One was vaccine driven. There was a mass onslaught of jabs in the USA immediately post WW1!

Events in the United States underline this interpretation. Even at this stage there was no ability to select, to purify or to deeply analyse the matter that was injected. It was still simply voodoo - however it was also rapidly becoming a major money spinner.

Where there's money, avarice congregates. Here, though, could also be used the tag of virtue – "we work to save you from these terrible ailments". "Wow", thought the politicians, "I want a slice of that". They could run on a ticket of aiding public health and fighting disease. And stuff their back pockets….What, no, I didn't say that. Did I? Oh, yes, so I did……[See also, in Book 1, "The Medicine Men and Doctor Angel".]

Most people who look carefully at the facts know and understand that death from all infectious diseases was dramatically falling as living conditions improved throughout the developed world – Europe, North America and better off regions everywhere. Never wanting to allow truth to overcome a good business model, the newly formed three part alliance of medical practitioners, pharmaceutical manufacturers and politicians drew in several other sectors of society – namely the information media of advertising and news – to instil fear and respect into the public. This became pretty much universal – vaccines fought off the common enemy of disease.

Jenner's evil concoction and lethal method of administration were dim and distant memory and never discussed as doctors or nurses injected, using sterile syringes a tiny dose of a sterilised extract …….from the same pus as ever.

By this time immunological sciences were developing – indeed, my father was a research immunologist for the UK Agricultural Research Council. Proteins in the bloodstream were found to derive from an earlier encounter with a particular infection – named "antibodies". They were shown to be individually tailored to the infectious organism, now christened an "antigen". A vaccine, of pus or extract from pus or, even, quite a pure separated fraction of puss later cultured on as "the infectious agent", [this is a long and quite well covered elsewhere saga, which I'll now put to one side] yes this vaccine can be shown to give rise to circulating antibodies to the components of said jab. The pus extract, that is. This is all the work, as I've described and discussed in The Immunobiology Section, in Book 1, of B-cells and, then, T-cells and hypermutation and Stuff Like That!

But they are still injecting purified extract of pus and still doing a Jenner. And still denying there is any significant collateral damage and still suggesting that all the antibodies are needle sharp weapons the body then has to defend itself in future against evil attacks by the specific dastardly terrorist viral particles.

OK, today they use expensive laboratories equipped with all manner of analytical equipment. I've used them and I know they are indeed powerful. But they cannot change the nature of the process. Discussing computer information processing they say: "You put rubbish in, you get rubbish out" and, classically it is noted "You cannot make a silk purse out of a sow's ear". (You can make a very cross pig, tho'!). They have attempted to ink in a back story in order to legitimise the whole vaccine industry and, frankly, it just does not wash. Vaccination is just one vast, ongoing fraud.

In itself that's bad enough but, then again there's a whole lot else in today's world which rests on very dodgy foundations, but what I just cannot accept is the denial of the collateral damage and the treatment of those who complain. Unnecessary death, illness and lifetime disabilities abound which are directly resultant from this process. Simply **unacceptable.**

5 - "So, it's like a great big conspiracy, is it?"

December 4th, 2014.

And

"All those people, all over the World, in so many locations and disciplines all plotting together to make sure that vaccines are still used and that reports of negative outcomes are ignored?"

And

"Come on, I can't buy that, the NHS, all those professionals, they wouldn't allow this to happen. They want the best for their patients, for their public".

So:

How do you bring on board those who have no direct involvement – or realisation of involvement – and who have always lived assuming you need vaccines to keep you safe? Does it matter to these people? Need one open their eyes?

Because we are straight back deep into "Conspiracy Theory" and the murky areas such discussions delve into.

Shall I perform a CT scan?

Meep, meep, zip, zap, zoodle.

OK, scan done and I have to report the good news that utterly nothing showed up. That's right, there is no trace of a conspiracy here.

So what is going on? How is the lid kept on and the spread of malcontentedness so rarely reported. How is the show kept on the road despite all the indicators that it is just one huge big misunderstanding?

Every piece I write is a stand alone, a separate chapter, but based on an amount of my previous utterances. I have previously looked at writings to online discussion forums to depict the manner in which Joe Public views the subject. I do not visit such sites any longer simply because the aggression of the pro-vaccination posters , or numbers of them, has grown to frankly unacceptable levels. Death and personal injury threats, coarse insults and out and out denial of chronicled facts. These are a group with an agenda.

"A group? Come on, I thought you said there was no conspiracy."

Conspiracy: *A secret plan or agreement to carry out an illegal or harmful act, especially with political motivation, a plot.*
Theory: *A plan formulated in the mind only, speculation.*

So, no theory here. This is actual. This happens. These people are paid or groomed or a bit of both. I once happened upon two of them, messaging each other about a newly posted contentious and pro-vaccination article on "Huffington Post". They were discussing what they would say to the anti vaccination element "when they find this article", suggesting ways to make them squirm and insults they could sling. A bizarre sport and as far from rational discourse as one could get.

You see there are many structures, many long established hierarchies and networks being defended here:

1. The reputation of the medical practitioners. For two centuries they have toasted the benefits of vaccines and, unlike so many other false medical icons they've kept Jenner right up there on a pedestal and cannot start to think that maybe he was an insane charlatan, any of whose patients were lucky to get away from him alive.
2. The vast profits these days generated for the pharmaceutical industry by the global penetration of the practice. Now grown from two or three jabs per child and far from comprehensive coverage up to dozens given out under near military planning and with obvious aims to build schedules of adult vaccines as well – flu jabs, for example. Yes, they invest much capital in product development but afterwards jack up the stakes and use a weird

social blackmail to drive governments to purchase the commodity and so make sure it is used.

3. Which brings me back to politicians. Oh my, really, if ever a theme was made for the candidate running for office, it is this one: "Yes and, of course I will put health at the very top of my agenda. New vaccines will be brought in to tackle all your ailments. There may even be one soon to make him want to go to school, smile smile, chuckle chuckle, you have to vote for me." The admission, as Edwina Currie had to, that a jab she'd recommended for one shot giving "A lifelong protection from these three ailments" [measles, mumps and rubella from MMR], actually did emphatically not achieve such effectively ruined her front bench career.

4. The Economy. Yeah, the whole schmuck would topple. NHS would be devastated, the government in tatters, Big Pharm out on the pavement begging when this all got to court. When the full impact proverbially hits the legendary fast rotating cooling device then there will be such an almighty schplatt and so much compensation to be paid out there will be total bankruptcy. It must be noted here that the United States Government AND the industry recognised this looming pitfall a good three decades ago when they set up a no blame scheme for Big Pharm. Even if you have 100% proof of cause and effect – your jab killed my child/precipitated his autism/gave him SSPE – you cannot sue in the States for compensation. Instead you go to the Government who give you hell and, if they give you anything, do not admit flaws and make you sign confidentiality clauses to keep the story suppressed.

5. Big Pharm impact on the media through advertising or its removal. Print or otherwise distribute damaging information and you go bust. If "New Scientist" print a full investigation on the Collateral Damage arising from vaccination or even start to question the supposed science of the industry then how much longer would they remain "the leading popular journal of science"? One week. Simply that – they would be dead.

6. Friend of mine is a research geneticist who has input into gene modification and its use in pharmaceutical products, including vaccines. He's talked through these issues with me and essentially says "Yes, you are absolutely right and I agree with you totally. I couldn't say so, of course. I'd lose my research funding and my job soon after." No, really, that is exactly what he

says. He goes on "Well, we've got to get the kids [now in their teens]through their education and pay for the house." And the holidays and the general good, middle class lifestyle. No problem just so long as you keep doing what you're told.........

7. And there are many other disciplines also deeply knitted into the fabric. Police – who have to arrest the parents for "shaken baby syndrome" or simply "cot death". It's just a job, they perform as instructed but, deep down, subtle educational processes obviate the policeman's need to investigate. Given the cause – the demonic parent(s) – he looks no further and does not question the responsible medic's input. Job done.

8. Then there's judges who sycophantically hang on the every word of their "expert witnesses" such as Dr "He Might Now Be A Doctor Again But My How That Shows Corruption In The System" Roy Meadow but rule as invalid or irrelevant testimony by the grieving and now indicted parents. Such as the late Sally Clarke, a lawyer who Meadow effectively condemned to death with his lies, deceits and misdirection. Not content with killing the children so the system then finished off their mother, after convicting her of their death and imprisoning her. After her eventual release – on a convenient technicality and not an "Oh my goodness we were so wrong – please accept our most humble apologies and regrets and let us do all we can to help you rebuild your shattered life". No, they just let her out and she had lost all will to live. Hell, you would have too – think about it! (Section 3(2) in Book 1)

9. And there are social workers – shall I go there? Bless them, hard working souls that they are. Working all hours in oh such dangerous communities. And, my, there's all that paperwork. And meetings, case studies and, well, you have to liaise with SO MANY other interested parties and agendae. And what with Maisie having trouble at school and I just could not get the car to start and I had to pop home to see to the plumber and then go and get my flu jab. How could they ever even question their visit to Ms X to find evidence of her mistreating her six month old baby? How could they be expected to doubt the GP's report?

I could but won't go further. There is no over-riding conspiracy. It is not needed for the system is self monitoring, self correcting, even. It does require a lot of hard work from a lot of committed folk to keep the show on the road and the major groups involved – that's 1 and 2 above – provide all the drive for that.

But they are not "conspiring" – they are just carrying forward their faith, their belief in this system.

Yes, I do believe Big Pharm employs – uses – minions to post online and perform a host of other public interactions. I mean there's the Grim Saachis of this world who take instruction, gross fees and alter people's perceptions. Hi Tec Big Pharm have product to sell and as the rewards are so enormous it is a market they need to maintain. They are also survival and profit driven and their vision flows through a very constricting tunnel.

So different companies sell flu jabs, measles jabs and so on. In competition with each other but all assume the products work as sold. They develop new product to fill new perceived need. Yeah, of course they create need, they push dangers of infection. So measles rose from common rite of passage two weeks in bed with comics to catastrophically life threatening infection we must all avoid ever contracting. B*llsh*t. It's only any danger to the malnourished and poverty stricken third world kids who don't die of measles – they die of that poverty.

Big Pharm also have near total control on the publishing of research evidence. In fact, they dictate what research is ever carried out, by and large. If they don't think your proposals will assist or grow their business why should they fund it or even smile upon it in any way? And they'll tell their friends what they think. Another friend, in the US, works on autism and deep genetic "defects" that can be examined. On her blog she discusses sundry such details as well as related issues. With such overwhelming faith she describes minute analyses of supposed errors in coding fragments of DNA, hypothesising how such tiny variations in order and selection of nucleic acid sequences could be translated into observed behaviour modifications in autistic individuals. (Section 4 (5) in Book 1).

There is no way such correlations can be made and she knows it but still goes back to her lab bench to churn out more such drivel. Why? Because it is funded, of course. Who by? Oh, come on, join the dots………

Which just leaves the medics, bedrock of the health service, who will do their utmost to care for and assist their public/clients/……. UK GP practices are paid a bonus, a fee, for reaching 90% of their infants vaccinated each year, and similar "encouragements are found worldwide eg to US Paediatricians practices. They are happy to do this **because they know it is a good thing**. Why should they be any different from the general public. They have grown up with the belief that vaccines keep people safe. They, themselves, will have been multiply vaccinated as kids and as students. Some must even carry vaccine damage.

Now that would be an interesting study! [ASD is a spectrum and the "good end" has been well chronicled as containing very mentally retentive, if rather narrowly focussed individuals. Excellent as medics?]

So why do they not respond more sympathetically when confronted with cases of obvious vaccine damage – cot death, asthma, allergies, hyperactivity, autistic spectrum disorders etc, etc.

Like the doctor, the GP, whose own son became progressively more and more autistic the more MMR jabs his dad gave him. Challenged on this the poor benighted doctor said it was pure coincidence and would have happened anyway. This, I believe, is very common although this particular guy was making a side career out of the issue, which I found particularly alarming. Sort of "NO. That cannot be true. I don't even want to think about it. I won't think about it.
Because it is not true. Whew, glad I worked that one out."

Also known as "Cognitive dissonance"!

6 - <u>Jenner's application for funding from the Medical Research Foundation and the Pearly Gates' Foundation.</u>

February 17th, 2015.

Abstract: The applicant intends to develop a natural extract from surface tissue lesions of Bovine species, post the viral infection known colloquially as "Cow Pox". This extract he postulates will be effective as a prophylactic medication to be used to allow recipients to naturally fight off infection by the often damaging or lethal infection of humans, known as "The Smallpox".

Method: The initial study proposed will utilise just one subject, the young son of a landworker near to the applicant's residence in Gloucestershire, in the British Isles. The extraction process, from the lesions on the body of the cow will entail scraping with a reasonably sharp instrument the pus which will, at that time, still be contained within said lesions and collecting it in a glass dish.

Some time later, maybe days, maybe weeks, the young experimental subject will be brought to the experimenter's premises where he will be prepared for the interaction. His arm will be bared and the experimenter, a licensed Doctor, of course, will use a knife to cut a deep slit into said arm. With reasonable haste into this wound will be introduced a portion of the bovine pus extract, collected previously. Portion size will not be recorded – suffice to note that "enough" will be introduced.

The arm will then be bound in bandage both to stem the bleeding and in order to keep the pus within the bleeding tissue.

Some time later the subject will be brought back to the premises of the experimenter where a similar extract of post-infective pus will once again be introduced into a new cut in the boy's arm. This time, though, the pus will have been collected from lesions on the skin of a smallpox sufferer. Living or dead need not be specified here but, for decency, we shall assume the former. Once again the bleeding will be stemmed by tight bandaging and the materials introduced kept within the boy's circulation.

Controls: There will be no control group for this limited, initial experiment – it is based only on whether the boy pulls through the process.

Results and anticipated outcomes: We are uncertain, of course, as to whether the child will survive this process but, if he does, we can create a whole new multinational industry on the back of it.

7 – A short note on 2015 Flu Provision – note my phrasing!

February 17th 2015.

Headline February 6th, 2015: "This year's flu jab *protecting only 3%*"

OK, so if only 3% of this winter's flu jab customers "were protected" that implies 97% of those so jabbed fell ill with "flu".

Is this "flu" further example of collateral damage?

Shall we estimate 10% of the population at large go down with a flu each winter?

Thus a massive additional 87% of vaccine recipients developed flu, so way over the population mean.

Coincidence?

No, I didn't think so either!

Statistical anomaly?

Ha, ha, ha!"

The lobby of activists opposed to vaccination have not even noticed this yet – despite my posting on the topic. Yet it is crucial as it so clearly exemplifies BOTH the problem with jabs – ie:

1. They do not work
2. They cause collateral damage

References from media:

1. http://www.independentnurse.co.uk/news/flu-vaccine-ineffective-againstcommon-strain/73646/
2. http://www.express.co.uk/life-style/health/556462/Flu-jab-blamed-for-risein-death-toll
3. http://www.channel4.com/news/flu-influenza-vaccine-ineffective-healthwinter-deaths
4. http://www.dailymail.co.uk/news/article-2941896/Flu-jab-waste-time-97patients-Vaccine-developed-year-ago-no-longer-matches-virus-mutatedmuch.html

8 - Wheat Belly-Up

March 17th, 2015

I have frequently praised this book, this campaign and many of its observations....

"But?"

Yes, indeed "but"!

BUT I have always felt that, for all his long catalogue of detailed references to a whole range of scientific literature, he has not thought the propositions through. Not well enough, anyway. He does not explain satisfactorily the whys of the situation, nor attempt to explain how human physiology is unable to handle many of the changes wrought about by recent alterations to lifestyles, including diet, medicine, housing and working lives.

"All in due course" I have felt until last week when I thought I'd take one of my discomforts to him for discussion.

On facebook March 10th 2015 I posted to the Wheat Belly site:

"I remain worried about definitions of Coeliac disease and why there are two types:
1   – the classical emaciated, unable to absorb nutrients, cured by GF diet
and
2   – the highly overweight, modern neo-coeliacs, who seem to absorb drastically the wheat derived glucose for diabetic obesity and again are cured by GF diet.

"How is this the same condition?"

Wheat Belly replied:
"Variable individual susceptibility to the opioid mind effects of gliadin-derived opiates and variation in bowel flora."

Hmm, does that translate into anything cogent? No, it does not!

So : **<u>Chris Hemmings</u>**
"…….I remain worried…….(continue as above).

"Me, I think the immune deficiency is triggered by childhood vaccination processes. Collateral attachment of immunogenic adjuvant chemicals in the jabs to wheat proteins in the blood stream derived from the individual's normal diet.
"There are many clear descriptions of the passage of wheat gliadins through the gut epithelium into the blood stream – even through into the milk of lactating mums, for that matter. As I say, cross reaction at the time of vaccination is, thus, a simple side reaction aka "collateral damage".

"What do you reckon?"

Wheat Belly – Professor William Davis, cardiologist of Milwaukee – did not come back either to explain his original offering or to answer my observations. All I got were a few random responses from others on the Wheat Belly site:
<u>Fred Dempster</u> "Why worry? Just don't eat wheat/grains/oats/corn and avoid excess sugar".
<u>Chris Hemmings</u> "I don't! Not, however, because I am coeliac but because, many moons ago, I chose to."
<u>Ron Gillespie</u> "Excellent point. I think I'll do that."
<u>Robert W. Rominger</u> "Like Fred said, your question is irrelevant. Forget about celiac. Forget about gluten,… wheat gluten (gliadin), corn gluten (zein), rice gluten (orzenin), oat gluten (avenin),…
"When you are grain free, you avoid all grains, all the nasty substances in all the grains, including each grains' gluten.

"The best thing anyone can do for their healthy (celiac or not) is to be grain free."

And there it ended. Yes, of course I agree that the individual wants a result, they want to lose weight and gain far better general health. They want to absorb nutrients and use them to build a new resilience, to restore their lives to the optimum state – if ever they had known it previously!

But, please, be interested in why the decline had set in. Wonder what had compromised your vitality. Look to avoid the problem for others into the future.

One day I may get back to Prof Davis. His observations on glycemic index and thus the devastating impact of ALL wheats on blood physiology, his delving into history and re-establishing F Curtis Dohan and his work on the impacts of wheat on the mentally ill, eg schizophrenia diagnoses, and his painstaking assemblage of so much published data to support his case are all very worthy of praise, and the results are clearly very impressive.

But it is not a complete picture and needs must be tied into all the other relevant descriptions, analyses and derivations – most of which you will find on this blog!

9 - The Voodoo Ponsi Scheme.

March 17th, 2015

The last few years have left us all tiredly aware of the Pyramid-Max mechanisms of financial skulduggery utilised by the heartless psychopaths of Wall Street and the City of London, and sundry copyists throughout the Global economy. Being able to keep a straight face whilst someone signs on the dotted line having been assured of the continuing upward trends of the product has been the most apparent qualification to rise up the corporate ladders in the drive to maximise incomes raked up by such companies and the individuals running them.

The Ponsi, whereby the only real income is from new sign-ups to the product placing substantial deposits of real cash – wherever derived! – and payouts are minimised by urging "reinvestment for even greater returns", the ponsi is derived from pyramid selling schemes, such as for soap, whereby those at the peak of the pyramid are made lavishly wealthy by the lowest levels – the recent entrants – of the scheme, none of whom can ever attain the rewards unless they enrol a complete pyramid of vendors beneath them.

Very, very few can possibly achieve such a goal, and then it would only be by "enslaving" others. Such schemes are essentially parasitic and the polar opposite of cooperative ventures.

So a "**Ponsi scheme** is a fraudulent investment operation where the operator, an individual or organization, pays returns to its investors from new capital paid to the operators by new investors, rather than from profit earned by the operator."

Voodoo ponsi? Well, this is how it struck me in conversation recently. There are several lines of intensification circulating through news and discussion group channels which demonstrate this.

Margaret started a chat:

"Well, we all know that vaccines are not about making money. They are about "preventing disease" and "saving lives," right? Have a look at the trailer for a new report on the Vaccine Technology Market forecasts to 2019.

[I could/should have added her reference here but didn't. The said piece was a slick marketing projection, painting the picture of a steep increase in business and profitability.]

"Nope, no money in vaccines for drug companies...." she added, with the very opposite implied!

"Tell me" , she continued, "if vaccines are so effective at "preventing diseases" then why is the financial forecast of the "vaccine technology market" so dependent on a "rising prevalence of diseases?"

"Were diseases so rapidly on the rise, as they [seemingly] are now, in the prevaccine era? Or, did we then have to deal only with the basic diseases that were around — you know, those ones that now are proclaimed to be "vaccine preventable?" How could it possibly be that new diseases are on the rise at such a rapid rate only after the religion of vaccinology has found itself so many new believers?"

Margaret pointed out, and here IS that reference, "In *Vaccine Technology Market by Types, Trends & Indication – 2019*, Prasaad Shelke, an SEO Expert in Pune, states that:

"The vaccine technology market is expected to reach $57,885.4 million by 2019 from $33,140.6 million in 2014, at a CAGR of 11.8%.

"Major factors driving growth of vaccine technology market include:

1 – Rising prevalence of diseases.
2 – Increasing government initiatives for expanding immunization across the globe.
3 – Increasing company investments in vaccine development.
4 – Rising initiatives by non-government organizations for vaccinations.

<u>Chris</u> These folk are quite scary. No, they are exceeding scary. They, of course, believe themselves to be a force for good……**That's** what I find most scary.

<u>Margaret</u> How about that calculated annual growth rate of 11.8%? !!! Wonder what the CAGR of new cases of autoimmune disorders among the vaccinated is …

<u>Chris</u> Exactly – you can see where/how they get their investors and then it's probably another Ponsi style business but with the twist being that the market collapses when the truth gets out, when the truth about the whole voodoo charade emerges.

Thus, one way to avoid this eventuality is, indeed, to invent new diseases and to heighten the purported risks from existing ones to the absurd stories that circulate in the media and journals. The recent "Ebolevent" being typical of this.

Collateral damage must clearly increase as a function of the number of jabs dispensed – either grand national totals or per individual. On the individual I feel it may well be an exponential curve. On national levels I'd need to talk with some epidemiologists. Obesity and diabetes in the states and probably in the UK are both clearly on such an exponential curve, as I have described previously.

Fascinating stuff, thanks <u>Margaret</u>.

<u>Chris</u> PS – On Page 8 in "i" today [March 14th, 2015] :

"Teens to get jab against aggressive meningitis".
All 14-18 year olds in England will be offered the jab
"cases have soared from 22 in 2009 to 117 in 2014, with 24 deaths in the last 2 years".
And, get this:
"a vaccine already exists"!

Sigh………Anyway, that's contextual for this Ponzi analysis:

1. They are constantly working to increase the number of recipients.
2. They are constantly having to increase the number of illnesses to be included
3. With a globally maxing population this is quite simple unless:
4. Negative feedback from recipients increases above a crucial level.
5. Which is why security of the delivery system is currently being so increased, verging on mandatory in many countries already.
6. And the publicised "dangers" of the illnesses are shouted so loud.

10 - <u>Akin to a bottle of alcohol for an alcoholic,</u>

the Government's Joint Committee on Vaccination and Immunisation (JCVI) and the NHS resolve to give an annual flu jab to all the obese in the UK. March 18th, 2015

Need I go on? Asked that before, have I not? Well, some of this is just so obvious and the official responses just so dumb.

Look, I've got to write this down. In the news today is the above decision. [For example:http://www.theguardian.com/society/2015/mar/18/morbidly-obeseengland-flu-jab%5DUsing the observation that obese people are most at risk of suffering flu badly they say that these folk could and should be given annual flu vaccinations.

William Davis would say, as would I, that a wheat free diet, low sugar, low processed food etc, would cure them. However we live in an era of freedom to set new norms. So they are not "obese", they are "differentially embodied" or some such assertion. What they are defines the norm – and HAS to be accepted.

So ASD, the autistic type designations, are termed "neurolinguistically variant" or some such escape clause. "They're not ill, just different" – and we have to adapt our interactions with them to reflect these differences.

So a broken leg is, of course, a broken leg, and all see the problems that ensue and help as and where required. The prognosis is usually good, if disruptive in the short term. Loss of a limb is far worse, as is loss of sight or hearing, because this brings about a permanent alteration to the individual's capabilities. We all can sympathise and understand such events and again work to ease problems as we can.

Now we are way, way beyond reference to race or colour as impacting on any individual's abilities. [OK, I'll exclude police and military from my "we".] I do recognise, however, conditions such as Down's Syndrome, where the individual has an extra chromosome in their genetic makeup. In such abilities are altered and, tho often sweet and charming individuals, their functionality is markedly reduced.

Autistic outcomes can be like this but are usually sadly introspective and hard to interact with. I needn't go through the staged descent and the clear collateral damage there is to the series of childhood jabs. That's a given, here! What I can say, tho', is that the end point reached is that where an altered outcome from that genetically programmed is reached and yet the attempt is being made to describe it as "a normal state – just different to that which you might expect" so "go with it, accept it and do not make critical comments" – as if talking about eye colour or skin colour or similar variant character.

Clearly they are where they are and all have to accept this – although I have such praise for those who steadfastly work to reverse the changes BACK TO THE NORM. [Folk like Shelly Tsforas in New York.] However it has to be regarded as a damaged state to obtain such restoration, to receive greater understanding from public services and, in due course, even reparations from the industry which caused the damage.

As I said, Dr William Davis – of "Wheat Belly" fame – would just say to the obese that they should go on the gluten free diet and, if they did, they would indeed stand to lose a considerable amount of weight. But this obesity in itself is so unnatural. So many, so obese just did not happen even 20 or 30 years ago and, as I've previously described, the many pressures on individuals today are all predicated upon the soaring levels of childhood jabs which can so readily lead to both auto immunity and, crucially, wheat sensitivity.

These are the modern Coeliacs as far as I can see. These are the modern obese. If we can stop their immune compromise – stop the childhood jabs – then I strongly suggest that wheats would be far less of a problem and obesity (with its twin ailment of diabetes) would rapidly decrease. Today's suggestion to give them annual flu jabs, notwithstanding the officially accepted figure of "3% effectiveness" this year for said jab (I kid you not!) would obviously compound the problem and not protect them at all.

I should PS that I do feel there must be both a reduction in overall food consumption and a redirection of food types eaten to a far wider range, irrespective of what I've just written. Modern supermarkets are full of so much and so expensive food of so low a nutritional value and, often, actually damaging to health. None of these issues are stand alone, clear, individual stories – there's always riders.

11 – Meningitis B a dead cert Money Maker

March 24th, 2015

Radio 4 of the British Broadcasting Corporation is a mouthpiece for The Government Line, the Voice of Big Business and the Emissions of Big Brother. Nothing they include is any surprise to me. I am always prepared, simply, to turn off the radio and fume. But something keeps telling me that I should listen to the news every morning. So I do – though generally still have to turn it off frequently!

Soon their morning news magazine Today, March 21st, 2015, the headline story was that: "A whole year has passed since the Joint Committee on Vaccines and Immunisation, JVCI, recommended the routine adoption for every child in the UK of the Meningitis B vaccine. Today that recommendation has still not been followed up and BABIES ARE STILL AT RISK OF THIS TERRIBLE DISEASE".

They went on to interview Distraught Mum who had sadly lost her two year old "to the disease" and was now having to spend hundreds of pounds to have her new baby given the jab four times, as prescribed, at at least £100 a go, privately. She was full of "why do I have to do this when it should be being given to every baby on the NHS?", crying on the listeners' shoulders.

The Presenter gave figures – "One in thirteen hundred will get the illness and of those one in ten WILL DIE". She rolled in a London Paediatrician who specialises in "getting children to be vaccinated against Meningitis B". He concurred that negotiations with manufacturer Glaxo Smith Kline were causing the delay.

There was talk of the costs. Using three or four jabs per infant, as recommended, comes to £400 or even £500 per head. Annually this is thus up to one billion pounds. Yeah, ANNUALLY THIS IS UP TO ONE BILLION POUNDS.

ANNUALLY. That's every year. Ad infinitum.

"Well, they have agreed to reduce the cost substantially from the list price ["list price"?- like they're a bottle of pop or a CD?] but the negotiations are ongoing. However, these delays have disastrous outcomes. Yes, the illness IS difficult to spot. Everyone knows about the slight rash and the glass test but really there's only flu-like symptoms so it's difficult to spot in the early stages." [No, seriously, that is EXACTLY what he said. No, seriously, I know it makes ABSOLUTELY no sense!] [He went on:] "And not only is it frequently deadly but also one in four survivors are damaged for life – they may lose a limb or there may be brain damage."

As I quoted at the very start of this set of investigations:

"Vaccines?

" They are the foundation of the modern health care system – they create customers for life".

No matter how short.

12 - When did you last see a robin sneeze?

10th May, 2015.

We carry colonies of trillions of commensal bacteria on our bodies' surfaces. That's on our skin, in our pulmonary tissues where air freely circulates and, of course, in our alimentary systems. Within our tissues, beyond those barriers we carry none. Plate out the blood sample, put it under a microscope and there are no colonies of bacteria. Take a swab of the cerebro-spinal fluid and do the same. No bacteria. Anywhere you look, within the body's "external walls" and bacterial presence is verboten. Simply this is because the gatekeepers, the clean up gangs and the passport control operators are just so damned good at their jobs. Nothing gets through for more than a millisecond.

Then there are virus.

What's their ecology?

Clearly they do get ingress beyond the above described security services. They do get to circulate – albeit at tremendously low concentrations – within body fluids and even ingress into tissue cells, where they can take over their running processes and redirect affairs to meet their own needs. It requires what may be described as a "deep-clean" to remove such invasive insubordination. The same fundamental truth exists, though – ALL such entities are to be removed and so the body's janitors essentially have a straightforward task.

Books like "IMMUNOBIOLOGY" by Janeway et al ("CD Rom inside") describe a whole range of innate immunity systems, with a range of specificities, that are used to cleanse such breaches in corporal security. Clearly, these are routine. Nature has, within its four billion years of pondering and development, allowed a balance between opportunist but wholly parasitic entities, virus, and the organisms within which they are biologically active, such as ourselves. It is a given that for a virus to bring about its host's death, it will itself be eradicated. This is, thus, not a good design and will not be selected for. It is, if you will, the viral equivalent of Kamikaze flying.

Thus a viral infection is keeping the virus in circulation, sure, but it is not an attack, a threat, to the host. On the contrary, it can be easily argued that stimulating the janitors into action, getting them out from their coffee rooms, playing cards all day, and giving them work to do, sharpens their skills and actually IMPROVES corporeal security.

In America they have huge amount of "Avian flu" in their caged flocks. If they detect it, it is curtains – for the whole shedful. Apparently they suffocate the birds by blowing in some kind of foam. Presumably like builders spray into cracks that sets solid. Land of the free, eh? Just don't cough!

Birds with bird flu presumably indeed gestate a whole lot of the virus to pass around the flock but why is this a problem? Perhaps because of the vastly overcrowded and frankly wholly un-natural manner they are being maintained. Bird flu in wild birds is not an issue – when did you last see a robin sneeze?

13 - Patches – we're depending on you

15th May 2015

Tobacco smoke is toxic, on a small, local scale it is the equivalent of the noxious plumes we see emanating from heavy, industrial chimney stacks. In days of yore – and it's not so long ago yet – entering a pub, cafe or the average office, as well as so many folks homes was to commit to breathing the accumulation of so many such effluents the impact was indeed industrial in scale.

The problem is abated now. The law has driven smoking into the more marginal areas and the actual process of smoking is so frowned upon that many, many have quit the process and see it as an abhorrent practice. Sadly, so many have also not got over their addiction, their metabolic dependency, for the nicotine they obtained from that smoke. As the widespread banning of smoking in public places set in, at the same time other ways to obtain the nicotine fix sprang up.

Latterly there has been a surge in "e-cigarettes", which seem to remain deeply addictive and a way to remodel the lost markets. Earlier, and still ongoing were "nicotine patches", being small, yeah, patches to stick to your skin which would release into your body over a period of hours a controlled dose of nicotine, such as you would have obtained by smoking during that period. Sold as a way to steadily reduce your nicotine intake, and addiction, they again had to fight off claims that they simply repositioned the addiction. Sure, it was more discrete and no smoke was required but when the patches ran out the addiction was still there. And so were cigarettes.

Me, I suffered passive smoking as my father slowly killed himself with all manner of tobacco products – cigarettes, cigars and pipe. I was never once even momentarily tempted by the pursuit and have never, ever tried it. Clearly there were so many things wrong with the process:

1. The fumes annoyed non smokers
2. The fumes lingered and, over time, stained rooms, curtains etc
3. The process was addictive

4. The evidence of a damage to smokers' physiology and the carcinogenicity of the smoke was overwhelming

5. The products were expensive – why burn money, why commit to unnecessary spending

6. In a world of food shortages why use so much rich land to grow such a damaging crop?

Thus the personal benefits of whatever the nicotine "hit" gave to the smoker were outweighed by far too strong a pile of disbenefits, which were carried by the whole population. Why carry that on your shoulders?

Why indeed! Peer pressure coupled with persuasive advertising, curiosity and the power of addiction. Why, there are also industries to support! Think of the growers – how can we take away their lucrative commodity crop? Then there's the factories where the product is processed and the vendors who bring it to the consuming public – all need the incomes.

An uneasy balance has thus been established wherein we know none of this market in needed but we can live with it so long as it does not intrude upon non smokers in any significant amount. Me, I'd tax the tobacco companies to fund the hospital and other charges arising from the damage their products cause and work to educate growers to eliminate tobacco from their portfolios in favour of beneficial crops – many more lucrative food crops, sure, but also products such as hemp or longer term tree crops.

Would that it were so simple for the products of Big Pharm! Would that their squirming was a sign that they, too, realise when to accept that, yes, their products are harmful and that they have a duty to assist in avoiding such outcomes. But, by a process of convergent evolution, they too have come up with the "patch". As I have described previously, a canny Australian realised that sub-cutaneous tissue is a "sweet zone" to find anti-invasive responses in host physiology. Yeah, it's a place we react strongly to antigen arrival. Gee, we all thought, what a surprise!

All over the body surfaces receiving direct contacts from the external environment are best served with such defensive/protective/maintenance facilities. It is, as could famously be stated, "a no-brainer". On your average castle, where do you station the guard? Yeah, the outer walls.

So he invented – and quickly patented – "The Vaccine Patch". Again, I have previously described this ["I wonder if he'll get these to stick?"], reporting that the patent had been sold on to serious Big Pharm company Merck. My thought was that they'd probably sit on it, to keep the market as they already had it. [Adopting to standard monopoly/oligopoly practice in buying out opposition or potential opposition.] Or they might hit the world with it as new, digital age, cut-out-the-middlemen product.

Far lower doses, very good antibody production and you really do not need any medical practitioners to stick a very slightly spiky patch onto your arm, do you? After all, have not ex-smokers been doing this for years?

Anyway, it seems Merck Industries probably sensed others in the market had similar schemes so they could not keep a lid on it – isn't the profit motive wonderful......? After less than two years, vaccine patches are so back on the table and going into development phases. I am sure it will not be long before you can go to your chemist for a roll of Patches for any of several dozen ailments – many we have not even heard of yet.

"Here is the 6 O'clock news. Reports are coming in of an outbreak of internal organ upset, IOU, sweeping through the Midlands and South Yorkshire. This terribly dangerous, debilitating and frequently fatal condition, caused by a particularly virulent strain of Escherichia coli colicoferens 999, has luckily been well researched by our beloved PPC, Patch Production Company, subsidiary of Mercoklinecomeagain, who have supplied not very expensive Patches to chemists up and down the region. NHS Prescriptions, plc, have a special BOGOF offer if you bring your kids in at the same time".

Yeah, you get the picture!

As I ended my original article on this subject, when the patch was still just a gleam in an Australian's eye, this process simply underlines how pointless vaccination is – in any form. These patches will house fragments of antigenic material together with adjuvanting materials and sundry preservative elements (these items will need a shelf life, obviously). They will be absorbed through the skin. As are countless antigenic items 24/7 for 365 days of the year. As happens also in the gut and in pulmonary tissue. It's what our bodies do. We are good at it. We do not need any help. That is why the sub-cutaneous cells are the sweet-zone – this is their job.

So hijack will be achieved through using the carrier chemicals and the same mismatching could still occur as happen today, precipitating the same collateral damage. The lower dose must be a blessing but, really, there is no need for any dose and all attention should be focussed on optimising the body responses to natural antigen exposure, not adding new, synthetic ones.

14 Jenner again

3rd May 2015

Yeah, I've been here a few times – but it remains crucial. When do vaccinologists see as the time that a barbaric practice, no better than primitive quasi religious and, as judged by today's standards, wholly damaging to anyone's physiology and continued survival, when did this practice become state of the art and essential to every human FOR their very survival?

Jenner was a charlatan and a rogue. His work killed people and it is very obvious to anyone today why this happened. I would challenge Paul Offit, for example, to apply smallpox pus to an open gash into his arm and then bandage the wound and wait a few weeks for developments. Ben Goldacre – you fancy attempting this or would you perhaps see it as "Bad Science"?

No, seriously, can you imagine what that festering pulp might have contained? A wide range of bacterial types, for sure, largely putrefying, plus fungal colonies and who knows what viral contaminations..........

And, over fifty years after Jenner, they brought out, in Edinburgh, I believe, the hypodermic needle to introduce their pus extract directly into the blood stream. No sterile technique – no need for it as no concept existed of "bacterial infection" to entail the need for sterile technique. As Dr Semmelweiss so clearly showed in a connected but different context. I guess I could also make reference to the state of street cleanliness – open sewers so often – and of the subsequent ENDEMIC bacterial contamination of drinking water.

Yeah, and quite soon afterwards Pasteur surfaced in Paris to embed the need to appreciate bacterial existence and to suggest that it was these that Jenner's practice worked upon and that the practice was sound. Accordingly, he continued to inject cultured extracts of pus into recipients' bloodstreams. Which still led to countless mortalities – "Quelle domage, il est mort, en depit de le vaccine". Denialism is not just a recent phenomenon.

You up for a dose of Pasteur's vaccines Paul? No? Whyever not – look, I can see Ben is up for it. See – or is that green colour on his face not a sign of enthusiasm? Oh well, perhaps what they laud as the success of Jenner's methodology, as updated by Pasteur still was not safe? Perhaps they recognise that it was still a killer voodoo faith, driven by the integrated self preservation mechanisms of a very ancient Guild of Medical Operatives. But they could not possibly say this. Then, of course, in the twentieth century they introduced another motive – the PROFIT motive.

In fact, quite rapidly in the twentieth century, as civilisation finally started filtering down to the masses, housing, nutrition and so on improved substantially and with it so the scourge of bacterial infection lessened on curves racing down to zero.

But, hey, the Medicine Men didn't want to ruin a very good business model with a bit of conflicting reality. They had to work to promote their products and prove their necessity to every living soul. Armed, by then, with sciences – microbiology, microscopy, immunology, psychology etc – but still using an ad hoc cocktail of infective pus with bitters like mercury and other adjuvanting chemicals to preserve the supposed operative ingredients.

The same "Tincture of TB" – a culture derived from a long dead woman lasted the century out and spread globally to vaccine manufacturers everywhere. It was the one, I am sure, that my 15 year old elder brother reacted to most severely. His arm became greatly swollen and septic around the injection site for a couple of weeks, leaving a long lasting scar. His four sibs, myself included, did not have to undergo this process as my parents realised the dangers!

So the process was still, in the 70s and 80s, the same process that Jenner had launched 200 years earlier. Still, I hold, just as barbaric. Still simply an assertion with no rational evidence to support it. Vaccinology is the holy scripture of this voodoo practice, performing black arts with high tech, highly expensive apparatus and laboratories. Faith based medicines are outlawed. So should this one be.

15 The biomedical ecology of three sugars – sucrose, glucose and fructose.

May 2015

OK, I sort of suggested the discipline – though I notice others use the descriptor now – so here is a very good use of it. What are the physiological interactions in the gut and in the bloodstream of these three. Obviously this has to include other inputs which provide the components – so, for example, the rapid boost in glucose after eating wheat carbohydrate – and rates of use of and/or disposal of all three.

Start to examine published works, blog investigations and even text book references and it becomes an exemplar of, well, the Tale of Six Blind men and the Elephant once more. "I can clearly see" they all write and go on to describe another non-integrated impression of the mechanisms working and of the outcomes of such. There is certainly an outpouring of certainties in the face of the interactions of biological subtleties. There is, it seems, to find definite answers to complex interactions.

"Fructose is a poison" is one such extreme. "No it's not" might well be another! Better would be "What set of physiological circumstances can drive fructose consumption to yield toxic outcomes?". It is a simple sugar, of the same formula but different structure to glucose and still occurs in two isomeric forms. Whereas glucose is used directly in the glycolysis and other energy generation processes, fructose has to be converted by enzymic action first. It does not trigger the all important insulin release into the blood stream and is also less readily absorbed from the intestine. Our most common dietary source of sugar, sucrose, is simply one glucose bonded to one fructose molecule. This bond is broken prior to intestinal absorption whereupon the glucose is very readily absorbed whilst the fructose is only slowly taken in, often in fact resulting in the excretion of a good percentage of this sugar.

The great rise in sucrose consumption in the post world war two, 1950s onwards era and the profound and hasty switch to "High fructose corn syrup" from the 1980s have raised obvious concerns about their driving obesity and diabetes, as well as a range of other associated conditions. As I have also chronicled, the much more recent contribution of Dr William Davis in bringing forward the role of the wheat grains in this pattern illustrates the source of rapid elevations in blood glucose levels. Palaeolithic diets, also widely promoted today, reach the same conclusions, albeit by a rather different route. Both describe the imbalances resultant in modern nutritional intake and both suggest that many, if not all, of us are consuming a constant essentially toxifying diet, whereby the sugars in our blood are constantly pushing against their homeostatic constraints and are chronically pushing emergency release metabolic pathways, such as glucose excretion in urine (diabetes) or fat deposition.

As William Davis points out, too great a concentration of glucose in the blood raises its osmotic pressure too high – water would be drained from cells in the tissues above such a level, causing drastic organ failures. Fructose obviously has the same impact. However, as fructose does not lead to insulin release, its blood level is far less well controlled, save by the speed of absorption from the gut, perhaps, or its utilisation in the liver, where it is metabolised. If modern far higher levels of dietary fructose, both in sucrose and in the corn syrup, lead to greater levels of fructose reaching the blood then the hepatic processing will be the limiting factor, fructose could "back up" and osmotic pressure increase independent of the glucose levels.

The ramifications are many and varied and the impacts often extreme. Nobody can argue that we eat a diet even close to that of our ancestors and so our evolutionarily established mechanisms cannot be being utilised optimally. Surely we can and do naturally digest fruits all of which have a high fructose content. Glucose, if anything, we would formerly have consumed far less of. Why there is no "fructose-insulin" is an interesting question which I do not hear asked elsewhere.

This topic seems to have overlaps, as I have previously described, into the central arena of these investigations. However I suppose I use it here as a reminder of the interconnectivity of physiological systems – as the global warmonger Donald Rumsfeldt noted in a probably unique moment of vision "There are things we know, there are things we know that we don't know AND there are things we don't know that we don't know". Such are these collateral damages, imbalances and knock-on effects.

16 Conclusion

May 2015

So here endeth a collection of four years of musings, discussions, researches, analyses and distillations. Literary impressionism, maybe, painting a host of images and reflections in trying to get to the heart and soul of this core element of modern living. And finding, of course, that a broad array of seemingly unrelated areas are drawn into these pictures and that the sum is far greater than the parts except, and this is the essential point, there IS the one precipitating factor.

Professor Paul Shattock, when I talked with him after one of his lectures ten years ago, reacted strongly at my objection to the defence of the MMR campaign. I'd said "But the MMR cannot be treated as a single entity. It is not the MMR that causes autism – it is Vaccination. MMR is often the final precipitator but that is because it is one of the later jabs. Much of the damage has already been done – as collateral damage to earlier inoculations, like DPT, Meningitis and so on".

Paul said "No, you cannot say that. They would laugh you out of court. Any individual prescription has listed its possible side effects. Different chemicals thus have different side effects and cannot be considered together".

Of course I again objected strongly. There are so many analogies one can use to describe the accruement of damage on a single system by a range of different forces. Erosion of a cliff by wind, waves, frost and rain for example but, in this screed, the story of blood sugar is a good exemplar. A whole range of modern foodstuffs raise blood glucose levels – not just "sugar" but, for example, as we saw, wheat grains raise levels even faster and are so profoundly obesogenic.

But he would not move. All jabs must be considered as separate factors and not as contributors to the greater physiology of the recipient. Such shallow and wholly indefensible thinking. Maybe not Paul's, maybe "industry standard", maybe "the academic norm" but where was the joined up thinking, where was the systems approach? I felt he, a pharmacologist, was uneasy but had to hold the line. He, as the father of an autistic child, I felt totally supported me- but this man was muzzled, despite his brave work in pushing the case of this obviously increasing sector of society, of which his son is one.

The name "Autism" itself is, anyway, a catch all and includes a range of altered parameters. The range, and so the diagnosis, varies over countries and moreso between countries. Here, in the UK, vaccine damage has long been spoken of, from sudden death precipitation through to a wide range of lesser, generally chronic outcomes. Read the medicaments' leaflets – they list them all as "rare outcomes" and contraindications. Autism, although first described as a condition in the 1940s, did not come into the picture. Certainly, anyway, it was a very rare condition.

In the 1970s there was a long public outcry due to collateral damage arising from the Whooping Cough vaccine and numbers of takers dropped radically. Eventually this was played down, calmed down, pasted over etc and fell from the news. Then, the 1980s saw rising numbers of those described and diagnosed as Autistic. Conditions such as asthma, allergies, eczema, hyperactivity or ADHD, colics and a number of others were also increasing but autism, as a clearly chronicled series of regressions, where the young children had sudden declines from previously obtained abilities, the milestones of childhood, came to grab the headlines. Time and time again parents recounted how these declines were immediately preceded by the latest of their childhood jabs. More often than not, but absolutely not exclusively, this later jab was the MMR, hence the link between MMR and Autism was easily made.

As we have also seen, this link is so readily disproved by large population studies:

1. Because all jabs have physiological impact including generic collateral damage.
2. Autism is an essentially vague in definition, including a range of disabilities, and is an ongoing progression, over a period of several years.

It is clear to me, and hopefully to you by now, that the unspoken shutdown on any discussion of the efficacy or the downsides of vaccines, leaves vast amounts of research unstarted. When the size of the vaccine pharmaceutical industry is looked at, and the weight of their product development and publicity budgets are examined this shutdown is clearly not due to a lack of interest. It is due to corporate, integrated strictures.

Strictures make structures and the whole industry is predicated against any question of the efficacy of the process. It is deemed that in being able to demonstrate within the bloodstream of a vaccine recipient that a protein created by the body has been triggered which reacts with the applied antigenic material, is a demonstration of the success of the process. Antibodies are being created to "combat" the antigen ie remove it from the bloodstream. It is NEVER possible to demonstrate such a reaction actually "fighting off infection" – the ASSUMPTION is made that, if and when required, this happens automatically. Anything demonstrated is, of course, *in vitro*.

Then population studies are used and epidemiologists have to show how, over twenty or thirty years, rates of infection have altered, how mortality has declined. As reviewed and discussed earlier, this is just not possible. Rates of mortality from all the classic childhood illnesses were in free-fall to zero before any of the jabs were introduced.

So as far as I can see, it has all been a deeply damaging, total waste of time.

Book 3
Talking of Voodoo

**Vaccines -
Philosophies, Nutrition, Medicines, Media
and many Discussions.**

Contents

56

1 - Battlefield Blues

September 22, 2011

Battlefield blues

I wake up each morning, and get into my range

I open up the windows,

And I see there is no change.

Then I light up my computer,

It hums just like before And

you know what it tells me?

Yeah the battle it still roars.

I got the no-change blues,

Yeah, I got the no-change blues

No matter how hard I try,

Those folk they just won't lose.

Well, I'm not really a gnarled old blues singer but I do think that the wars of words have dragged on for a good two hundred years and there really should be a move towards consensus by now. Instead there is less agreement than ever and the medical-industrial complex, MIC, establishment is simply attempting to crush out any signs of opposition to their supremacy. Total domination would be their only acceptable outcome, it seems, rather than a reasonable and wholly objective assessment of the multifarious and lifelong outcomes of this never tested hypothesis now used unquestioningly on countless millions of the global human and livestock populations.

The hypothesis is that injecting live bacteria/virus particles or their denatured derivative products into a potential sufferer's bloodstream allows his/her immune system to develop an ability to fight off that infection without suffering the disease symptoms in the future. Vaccination to instil immunity.

You couldn't make it up, really. It just sounds too far fetched to be true. Surely there have been controlled trials of these supposed medicaments, these prophylactics that are so freely offered to everyone. The truth is that there have never been such trials and the effectiveness of the process is just derived retrospectively and indirectly.

Twenty years later they will say "Look, numbers suffering that ailment decreased". They will not count "other conditions" which arose concomitantly nor will they consider decreased diagnosis of a condition. "You can't be suffering from that – you've been vaccinated against it".

Anyway, I recently was taken to task for stating in public that I was, indeed, anti-vaccination. The critique was that I should only be proSAFE-vaccination and not close out the possibility that such pharmaceutical constructs might be achieved one day. It wasn't clear whether s/he thought any such were available now. I answered that I saw the problem when maniacal pro-vaccinators dispensed vitriol and hatred against any and all who even questioned the process.

Yes, I have been the target of such attacks and it is most unsettling. They use the concept, the inane concept, of "herd-immunity" to thrust moral indignation and shame upon those who do not vaccinate. "You put my young child at risk by allowing the measles virus to continue to exist – my young child is too young to have been given the measles jab so far so if s/he catches measles then it's your fault and I'm gonna kill you if anything happens to my child".

Ouch. This is the visceral, Andy, this is where it hurts. There's no cool, clear and rational discussion to be had here. It's the same logic as found in the support of a football team. "Celtic?" "No, Rangers – an' you??"

Yeah, I've got all the science, all the logistical discussions. I understand biochemistry, immunology, genetics, bio-medicalecology, nutrition and bacteriology. I've studied them, researched in them and have built up a very clear derivation of illness, protection mechanisms and optimal health strategies. They are a detailed hybrid of all those disciplines and some others too. But it does not distil down to a simple, reassuring phrase or two to pacify such ingrained aggression – rational or otherwise.

Truth is that the aggression is of exactly the same intensity when it emanates from a highly educated medical practitioner. Highly and incorrectly I would suggest but it's the only mould they have. The doctor who vaccinated his own son who then became profoundly autistic is an obvious case. How could he ever say "My God, son, I did this to you"?

It used to be, in the mid nineties when we started this trip of discovery, that one would not directly say to a new parent "Don't have your child vaccinated". You would talk them through information sources, give your own interpretations and calmly leave the other to make up their own mind. I feel the same now but am wholly clear as to my own, personal conclusions. I think people who go through with the catalogue of jabs for their newborns are dumb.

However I clearly cannot have a complete grasp of all the possible ramifications of our integral homeostatic mechanisms. It would be wrong of me to be so openly critical of the established practice and not leave windows open for new enlightened methodologies to be introduced and for a vast range of health maximisation processes to emerge. I have many firm ideas on these fronts myself and long to pursue them towards the greater human good.

At this present moment my conclusions are that we are wasting so much energy on this eternal bickering. I grant that there are many noble goals apparent in the wish to eradicate illnesses but am more aware of the deceit in a system that continues to ignore the damage resultant from it. I can even see that to be more open will profoundly damage public faith in the operation but feel that this is a bullet that must be bitten, an issue that must be addressed in order that we can then progress to a far healthier population maintained at far lower cost in both the short term, through childhood illness, but also in the long term as so many degenerative conditions are eliminated.

We should establish a wholly objective trans-disciplinary centre of cognisance, of interpretation, back-research and progressive investigation, to pursue all the elusive outcomes of our clearly deficient current "immune system interventions" carried out to derive improved health outcomes in terms of resistance to certain infectious conditions derived from viral or bacterial proliferations leading to sickness (Yeah, "vaccination!").

Yes, I'd love to be the first Director and to help guide the project towards such beneficial outcomes as I have outlined. The approach is, indeed, revolutionary and many regimes may be overturned. I envisage that quite soon though we would develop the twin outcomes of greatly improved health, including the elimination of chronic conditions that have recently so proliferated in children and now adults, and also huge financial savings in terms of far lower pharmaceutical bills and far less loss of time due to chronic illness.

Shall I start the fund raising now? It's never been a particular strong point for me, raising capital, but the need is so apparent I'm sure that we can draw the funds together. Where shall we start the bidding? Anyone got a spare Institute we could use? Couple of dozen staff should suit pretty well, built up over a year or so, so we'll have to fund that, too. £1000000 for the first year – then it's self supporting.

2 - Beware Psociety's Determinista

November 10, 2011

I have grown to admire the capacity for exchange and development of ideas that the internet now provides. With the right people answers can be moulded by drawing from all within a group. Unlike a chat or a public meeting, you do not need to take minutes for every word is, of course, recorded and read by all participants. Or can be. You can tell how well or not this has happened and, just as in face to face conversation, you can return to a topic, re-emphasise it and expand on what you previously said. However there is a very good measure of the body of the subject and how it has matured.

And then you can edit the record afterwards to show it all more clearly! Well, in the heat of exchange mistakes are made, plus you can correct poorly phrased points. In the following I have done just that. Whole portions edited out, corrections where appropriate and a clear flow so distilled.

Alan set the following puzzle: *Do we have any free will at all? Is it really your choice whether you watch this documentary or not? It is about psychopathy, serial killers and the implications that science has on attributing moral responsibility. "Horizon – What makes us good or evil?"*
http://www.youtube.com/watch?v=u88lYs4FMTY&feature=relatedBBC
So I dutifully watched it again as I'd seen it on live telly some time earlier. I replied:

"So we're talking of the accelerator functions in the control section of an X-chromosomal gene and variability to the coding section of the functional gene. These, it seems, in some cases, can take up memory of childhood trauma and precipitate psychopathy. Then again, sometimes they don't."

And I brought in the esteemed Oxford Professors Ponting (Genomics) and Talbot (Neurology) [Section 1, (5) "Genetics subverted – by its establishment".] quoting: " [changes] are often private to each individual. This tells us that different parts of the human genome can be disrupted independently in people with a single disease: there are likely to be many dozens, possibly hundreds, of "autism genes" for example."

"And further it is: "extremely unlikely that there are single genes for major mental illnesses such as schizophrenia"

"Finally they added that these: "insights should ensure that unwarranted pronouncements of fault are not levelled at parents who produce anything other than a "normal" child."

"Put another way this says we all carry variant forms of many genes and the interactions of them in vivo are outrageously complicated and unpredictable. And then you're born! Or, again, you could say that life is a genetic illness.

"And, yes, Genetics is my first degree."

And it developed. Carlos was interested but baffled by the subject. I tried to encourage him:

"Hey Carlos you ever heard of Sickle Cell Anaemia, SCA, in the states. Only ever in people of African descent." It didn't work, but I wanted SCA in the discussion anyway. I was asked to explain:

"Sure. SCA is a simple genetic illness, causing malformation of the red blood cells – sickle shaped not round. In native African populations in USA and Europe it is a problem, a genetic illness caused, I think, by a single gene variant.

"Back in malarial areas of Africa, however, it has the effect of blocking colonisation by the malarial parasite in the blood. In other words it has profound advantage and the gene in naturally selected."

Zana, a medic, pointed out it bestowed resistance to only one malarial type which I will research. Could that be counter evolution? Quite possibly there are many other such skewed interactions – ad hoc solutions to otherwise impossible situations. Working answers, adopted and so selected for, despite their imperfections.

"I'll get back on the Warriors in a minute!" I added, reference to purported warrior genes in psychopaths' genomes. So:

"OK, Chromosomes are not strings of beads as there are control areas to moderate the actions of and access to the coded directions for gene products. Like imagine if all your computer files just opened out at random – there would be utter chaos!

"As I understand the mono amine oxidase A, MAOA, gene it's operation has a number of control codings whose product/signal is required for the expression of the gene. The "Warrior inheritance" lacks all but a small amount of this coding, hence the gene in potentially underexpressed. Especially in the male as it is coded on the X chromosome. [Females=XX, Males=XY, thus males have only one copy of the gene and control sequences but females have two]

"Which could result in build up of mono-amine by-products from neural operation or other subsidiary side reaction chains. Akin to muscles producing lactic acid if you don't pace yourself, precipitating a stitch."

Julia: Question – Does the mere fact that you can locate an inherited malformation, 'genetic' in that sense, necessarily lead one to conclude that complex forms of social behaviour are also genetically determined?

"Wow, Julia, I hope not! Does not my counter-genetical cynicism come over in my comments? Seriously, tho', I used SCA as a straightforward, long known inherited genotype which carries more than a simple coding and then went on to try to show other alleles [gene variants found at any particular location eg dominant and recessive forms] can be a lot more involved.

" So what we eat, where we live, who we socialize with, what medications we're given (eg vaccines!), the stresses we encounter, air pressure, lunar phase, breast feeding as a child, our ages and-so-onand-on all alter our physiology, our gene expression and, hence, our actions. Oh, and alcohol, of course, has major impacts here!

"Anyway, as the wise Oxford professors stated, there's probably several hundred genes for autism. Carefully and individually tailored, they would have it, to the one who develops the complaint. These academics are so far from the real world I cannot honestly say I understand why they are where they are. It's just part of the great disconnect we're living through! But that's another "thread" to start." *Julia: Ah.*

That's the general problem with biological determinism in general – but the scepticism of scientists is never reported to the public and yet an unproven theory is taken up as a commonplace fact – ho hum… Ho-hum indeed:

"Julia – Because it's eye catching and sells newspapers, but also because it attracts research funding. So they set up Holy Grails to quest. Promise the best possible outcome and "show" how well they are achieving this in opaque data quietly published behind glaring publicity of the goals aimed at. Get dumb journalists to parrot off what's been achieved in glowing tones of adulation! It's called the New Scientific Method.

"There's lots else but I think it's most apposite to say that at present our evolution is way behind keeping up with our recent environmental changes, which have been profound. This is crucial to this discussion"

"Alan, as Erin just said, there's far more to it than simply one gene being selected eg different allelic variants for MAOA, differing numbers of control elements. There are at least two variants of the MAOA gene itself unconnected to the Psychopathy issue. It IS interesting that the MAOA may be underproduced in times when there is high use of the neural pathways, so potentially creating this particular behaviour. It could be a toxicity, acting like alcohol, for example, as the byproduct accumulates. And there may be SCA type exchanges, too."

Alan: Chris, but is it not still a valid question, what evolutionary advantage does a specific gene have? Bare with me here – genetics is a complex subject about which I know next to nothing.

"Natural selection acts upon the whole organism. The recently late Prof Steven Jay Gould summed up his life's work as an/the evolutionary palaeontologistic theorist in a book I have in front of me now called "The Structure of Evolutionary Theory". It is 1433 pages long. Neo-Darwinism updated and rarely uncontested it encourages the inclusion of chance. I love it. Also, simply, "what's there at the time is good.""

"The Dawkins came to fame shouting about "The Selfish Gene", though in conversation he backtracked. The image sticks and has managed to almost personify genes, battling it out on the high sierras of life. An oak tree produces 10000 acorns each mast but, during its 100 plus years, only a handful ever grow to maturity. That's luck, not skill, so selection takes a long time and identity must be "fuzzy". The Warrior locus seems to have impact on human behaviour, especially males, when the enzyme may be sparsely produced. With luck, in another, say, 10000 years we may evolve this effect away.

"And I know that's not really an answer, Alan, more an atmosphere, an ether for emergence. Scoping, as they say."

This was in the air but not absorbed I felt and matters drifted. I came back:

"Alan if I can just clarify further – there is no warrior gene. Instead, at most, there is less expression of a particular gene whose housekeeping product then cannot keep up with the work provided in certain situations.

"This means that any actions arising cannot be selected in any meaningful way. It is part of the greater genetic profile which is successful and whose makeup is subject to chance compositional change, within as wide a range of parameters of content as is biologically sustainable.

"Oh and I must just emphasize my "like" of your antipathy for reductionism, Erin. Curse on modern society it is!"

This was after she'd posted "Reductionists, (smh) I wish they could at least admit they are just now getting the iceberg in their sights, rather than claiming to have found any sort of key."

Alan: Chris, thank you for clarifying. I don't think that invalidates the discussion of the evolutionary advantage of psychopathy – not yet anyway – so the simplistic references to particular genes in my above comment can simply be replaced with 'hereditary trait'. In the documentary it mentions (if I remember correctly) Lizzie Borden, a serial killer and I think it says something about there being many other murderers in her family line. This suggests that psychopathy may be hereditary, and unless someone can effectively dispute this I think the question of whether psychopathy/lack of empathy has an evolutionary advantage is still valid

Alan: Erin, I think you will find that plenty of psychopaths have children. From what I can gather they can be very clever at blending in to normal society – take the scientist who presents the research in the documentary. Maybe I am pushing the point too far, but I thought about this when I saw the doc two months ago and convinced myself at the time that lack of empathy can be explained by an ancient survival advantage that gets coded into and played out in our evolution – and no one has yet disabused me of this notion on this thread.

Oh dear, I'll have to work harder, I realised. OK:

"Venn diagrams. Overlapping spheres of influence. One "psychopath" with blue eyes. Another with brown. One with SCA, another without. One with an IQ of 150, another with IQ of 80.

"The marine cadets learn to be psychopathic in behaviour, the child sees its parent kill. It's not "I am a psychopath – I must keep myself blended in until I can do my dastardly act(s)" as psychopath is the label affixed retrospectively by psociety. And would not a family tendency to killing each other be a very non-inherited characteristic! "On the specific of lack of empathy I would imagine any advantage for this tendency is so recent – such as surviving a visit to a GP surgery(!) – that natural selection is yet to act on that sphere of activity. For a social animal empathy seems essential in a natural environment. That's so not what we have now."

Julia: But Alan there are two things that could be argued against this

-

1.   A warrior is not just 'ferocity' – nor are they 'triggered attackers' – war, in all the history I have read – from ancient accounts to modern, is about 'the art of war' i.e. it is not about the ferocity of force, it is about efficient use of force. It takes deliberation and ideological purpose.

2.   There were warriors called Berserkers – but they used a narcotic to induce ferocious violence – but, although the Romans had some in their army – they only ever used them as a moral tactic – no campaign was ever won in Roman history because of ferocious violence. Caesar said wars are won by the 'spade' rather than ferocity.

3.   The point about imagining a tribe some thousands of years ago having some evolutionary advantage from psychopathic traits, has no meaning from an evolutionary standpoint – since the time frame needs millions of years, and not thousands. If they were so useful, they still would be in recorded history, e.g. Homer, if we include oral traditions written down, which is some thousands of years old and they could be in surviving modern stone age cultures – but it's demonstrably untrue.

That's useful – Berserkers maybe analogous to American Marines. And the time scale is the right order, although thousands are fine by me – millions allow significant human evolution. Further clarification in order, tho' we're progressing better now. Right:

"Ferocity is not the attribute under discussion. It is a particular lack of remorse and/or empathy which may relate to underproduction of a particular housekeeping enzyme. Ferocity and aggression are par for the course, we've all got those, and empathy is actually harder to judge, nowadays, but I hope/reckon it's still there as I'm sure it always was!

"And no-one says there's a warrior gene. It does not exist, we're talking control *sequences only (as I wrote up above).*

Julia: Alan you ask what, if any, evolutionary advantage do the genetic traits that make up psychopathy have.

As I understand it, this is your primary question – but Alan this means you already accept that psychopathy is entirely and exclusively caused by genetics, and that therefore they 'must' have some evolutionary advantage.

1. There is not enough evidence to say that psychopathology is entirely genetic – in the way, say, hair colour is.

2. Not all traits that are inherited are 'advantageous' – there are 'genetic pathologies' that are just that – pathologies. Williams' syndrome is one – its traits are not selected because they are 'advantageous' as it is the pathological result of genetic mutation.

So, Alan, this question must be answered first – is psychopathy not simply a biological pathology like other biological pathologies, e.g. Williams syndrome?

Good interpretation although I am in need of research into this newly established syndrome, also known as "Elfism". Also she is close to defining Psychopathy as a specific genetic illness. Anyway:

Alan: Julia, yes, now I understand. I was assuming that psychopathy was caused by genetics. I did admit somewhere in the thread to being almost completely ignorant of genetics in my appeal to Chris to indulge me.

1. OK, accepted.

2. And yes, I was also assuming that inherited traits have had in the past some advantage.

I am grateful to you for unpicking my unstated assumptions.

So now, how to answer the question: is psychopathy not simply a biological pathology like other biological pathologies like Williams' syndrome?

Alan: *Although the 'not' in your question implies that you already think that psychopathy IS a biological pathology. And I would guess that Chris agrees with you (didn't he say something like 'life is a genetic illness'?)*

My turn again. Nearing resolution.

"I did say that, Alan, though it sounds better in context! I also used SCA to support the idea of unexpected outcomes in gene expression. MAOA is even more involved, with benefits and disbenefits and probably a lot we are as yet unaware of.

"Be careful of labelling ANY genetic condition as "simply biological pathology"."

Then Erin stepped in

Erin: *I see the conversation is winding down but since I spent time thinking about this today I'll share what I came up with. Survival advantages of psychopathy:*

1. Keen observation of environment

2. Very difficult to catch them off guard

3. The slightest hints at intricacies in social structure are analyzed

4. Points of weakness tabulated

5. Because of their observation and analysis they are able to adapt into nearly any social environment for a short duration.

6. Charisma – ability to swindle

7. Ability to sow their oats and move on without guilt (obviously I'm speaking of males)

8. Adept at dog eat dog dealings.

I don't know Alan, it's kind of scary how many advantages I saw once I started thinking about it. From a lone wolf all the way to a corporate leader the list could get quite long. Although I still have to stretch for the family man/woman!

Alan: *So it seems that we are at the point where we can say either that psychopathy is inherited and has continued as a human 'condition' because it confers a survival advantage, or that it is just a pathology. My hunch is still the former although I lament my inability to convince anybody but myself – lol*

Now hold on Alan. This is not the end yet, because:

"You inherit aspects of your physiology which can in certain circumstances allow "psychopathic" characteristics to be demonstrated. Life patterns and environment can emphasize or intensify this but these patterns can also be induced in more "normal" physiologies (eg marines, Berserkers)

"I think your list is a bit speculative, Erin, although I'm sure that the Occupy movement will have "Psychopath" in their lexicon for describing the Wall Street Criminals!

"You know, Alan, there is no need to make your choice as it is part of the personality spectrum and, as such, neither pathology nor simple deterministic genetic condition. There's a kind of "no turning back" aspect to one's development through life and modern society seems to encourage avenues to be taken which promote "psychopathic" tendencies. Perhaps we should look at this aspect more."

Pause while took Tom to college and chewed over some more. Then:

"Is there a gene for musicianship, is there a gene for sportsmanship, is there a gene for genius? The answer is no, but within the inheritance are features which may assist in such progression. Follow that path and there seems to be positive reinforcement – "habit" even. We should take genetic labelling no further or we become determinista and encourage a New Eugenics."

Julia: Hear, hear Chris – I whole heartedly concur.

Wow. Was I pleased to get there. Two days of teasing out our brains and I'd got it all saved to edit at leisure. This is the future of journalism. It's also great evolutionary genetics – in more ways than one, I guess! I knew Erin would come back on my earlier reference, so enjoyed the last tidy up:

Erin: Chris – Yes it's speculative. I was trying to step into the mind of a psychopath based on what I have read. A simple mind game to sidetrack the logical debate and try to see right to the heart of what Alan was saying. If there are survival advantages, they MAY trickle backwards to genetics in SOME way. Just because genes are not the absolute answer to personality does not mean we should disregard their role completely.

"There's a kind of "no turning back" aspect to one's development through life". Do you really feel this way or am I taking it too far out of context?

" No, that's cool, Erin. I love speculation and I love your description, too. Kind of "method discussion", Marlon Brando style! My thesis is to take genetics off its present day pedestal and remove its accumulated fatalist/determinist streak. Modern use is horribly, yes, reductionist!

"Obviously our genetic programmes are crucial to the fantastic array of physiological functions we constantly run and to maintaining the dynamic equilibria amongst the vast numbers of variables, not least in our relationships with all other organisms we encounter both large and terribly small.

"We've all got that system pretty much adequate for all we attempt to do. Not "perfect", just adequate, sufficient and allowing fuzz and blur, drift. But I'm always in awe at the complexity and beauty of the myriad interactions. Chromosome pairing and duplication then division. And it just does it. Phenomenal.

" And yes, I do think there's a strong element of no turning back. Aging, obviously, but habits, skills, thought processes, ways you react, that sort of thing. I think it is quite difficult to unlearn especially if you don't realise the commitment you've made. And your physiology does the same sort of thing."

New words:

1. Determinista – Believers in the cult of pre-ordination. Genetically ordered fatalism. Akin to other theistic adherents.

2. Psociety – overly analysed social structure.

3. smh – Shaking my head.

3 - Just because.

November 16, 2011

There's the long, reactive, fully editorialised version. And then there is neat, pure posted of what was. Just because:

1. "I fail to see how you can 'cure' disorders such as depression by rubbing two sticks together whilst sitting under a yam tree".

-Well, it depends where the yam tree is but I'm fairly sure it'd make me happy. Remembering:

2. ".....people I know who work in psychiatry do not take the pharmaceutical route as their first option."

-Perhaps because the alternative of talking is preferable! Nonetheless, pharmaconarcotics and the like, "biochemical coshes", are so commonly used. I taught horticulture at a bail hostel for a couple of years and the bulk of the inmates were given handfuls of such tablets on a daily basis by their probation staff, as medically advised. Kids who'd got drunk or stolen to support addiction or just stolen for income.

3. "Alternative medicine is based on 'inductive fallacies."

-That's what one must term a damning, unsupported generalisation. Amusingly, it is itself an inductive fallacy.

4. "If there are logical fallacies in science it creates 'bad science."

-No, it is not science at all. Obviously. Ben Goldacre, bless his cotton socks, is probably himself an inductive fallacy. You had better ask him.

5. "I see nothing in the efforts of those who work in medicine that makes me conclude that they are conducting errors of induction."

-This is a statement akin to Horatio Nelson's so famous "I see no ships" with telescope fixed to his blind eye. Shall we talk of evidence based medicine, essentially prescription by the mean which, by definition is frequently the incorrect outcome? Or, perhaps I should expand on the archaic practice of vaccination?

6. "A bad science would be dogmatic. Medicine is not."

-Ask Andy Wakefield! It has to be – it has a reputation to lose. Not to speak of zillions in damages. Medicine is profoundly, overpoweringly dogmatic.

7. " However alternative medicine never alters it's fallacious conclusions."

- Now the conclusions are fallacious as well! Now where's the logic – if you say the assumptions are fallacious you must stop there!

8. "If it could be inductively proven to me that depression could be cured by foot massage – I'd advise a depressive to remove his/her shoes."

- "Inductive proof" pales when compared with witnessing the actuality, which I'm sure Alan and Erin could provide. Me too, only I've only done a weekend course run by another of my friends. It's the same principle as shown in the "polarity" film. Most of it is plain common sense, but yoga, acupuncture, massage are all profoundly beneficial and all stress diet and lifestyle as inclusive to the therapy and ongoing regeneration. Conventional medicine used to know this, too.

4 - How's this for Dogmatism?

November 28, 2011

"How can we demonstrate or exemplify the quality of good science?" Elizabeth had asked, elsewhere, but I felt it needed promotion to a separate discussion.

Me: Good scientists: Archimedes, Galileo, Newton, Darwin, Andy Wakefield

Moderate scientists: Mendel, Watson &Crick

Poor scientists: I'll fill this one demain as I've cut too many logs today and must go sleep!

Elizabeth: How do you measure the relative virtue of the scientists ?

Carlos: Two questions: 1) Is the person's knowledge and understanding of the nature of things reliable? 2) Can the person be relied upon to always do things in close cooperation with the nature of things?

Me: Hi, there's three good questions:

1 – Elizabeth – Virtuous scientists, en mon avis, do not shy from truth although it may be unpopular and present matters as they actually are. So Mendel, although he's the father of genetical theory, cooked his data to demonstrate his theories – very bad habit. Watson and Crick, first showed us structure to DNA but stole crucial data from Rosalind Franklin.

2 – Carlos – I feel their manner of approach says a lot about their reliability. Gallileo was fiercely attacked by the Catholic Inquisition to recant his science, but stuck to his support of Copernicus' heliocentricity theories. Andy Wakefield has recently received

similar treatment for criticising the current medical hierarchy – and was cast out by the GMC "Star Chamber"

3 - If they are 1 & 2 then I would expect a "Yes" here.

Then came Dogmatix:

Julia: Science is the application of a method. Of induction. Good science means nothing more than the application of induction to questions that are truly inductive. Bad science results from applying induction to questions that are not inductive. Pseudo-science is the application of fallacious logic to questions that are inductive.

To use the term 'good' as in virtue – does not mean 'method' – it means "what are the practical consequences" of an proven inference. That needs to be framed "what good does science do?" not "is this good science". The linguistic context is everything.

Just to make a point about that and about 'definition'

"All men are mortal." is not a definition. "Socrates is a man." is not a definition, it's a description. Therefore the suggestion that I cannot conclude "Ergo Socrates is mortal" from premises that are not definitions is absurd.

Even from the logico-linguistic point of view – definitions have no baring on sense. It is linguist context that changes or give the sense to the terms of a statement. Even terms that are synonyms can become antonyms simply by changing the linguistic context. e.g. Vision and sight are synonyms. They can also define the same thing. But there is a difference between saying "Barbra you're a vision." and "Barbra you're a sight." – the synonyms have become antonyms in this linguistic context.

Me: Julia, I much prefer deduction. Induction is what I do with copper coils.

Julia: As bad science must…

Sigh, how stupid is she? Does she take that as a logical or even humorous comment?

Me: Perhaps you never have but I imagine the principle is the same. Deduction is drawing out whereas induction is putting in, bringing to being, surely.

Julia: Not in logic

RE: The problem of induction:

http://plato.stanford.edu/entries/induction-problem/

OK, when given such a reference I will look at it, and so I did just that. It was a Stanford University philosopher, giving a rundown/intro to induction and deduction before going into reams of obscure calculations, pure mathematics and other banal, impenetrable sophistry. Of course it shows my analogy with copper wire is useful. Given this situation, they say what should occur. Take copper wire into magnetic field and electric current will be generated. This magnet or that magnet, this field or that field etc. From this they generalize to provide "Laws".

Me: The point about science is not to predict your results but to deduce from what you have seen any rules or techniques that may be applied in other situations. So you examine an emerald, do its chemistry, its physics and write it up. Future gems can be compared with your data and decisions be made as to whether they fall within your criteria, whether these should be altered and, indeed, whether this new gem is an emerald.

Or, as Stanford says:

"There can be no deductive justification for induction. Any inductive justification of induction would, on the other hand, be circular."
That'll do for me!

Elizabeth: Baconian induction is the method used by modern science.

Julia: I can only infer that if you use the following "There can be no deductive justification for induction. Any inductive justification of induction would, on the other hand, be circular." to as a definition of scientific method – you have not understood the statement. *Que, Julia, que??*

It means there is deduction that one can make to justify the inferences of the type you describe. Nor is there any inductive inference one can make for those kinds of inferences.

Que, encore, he bilingually interjects.

If you say "That'll do for me." then you deny that you can make the kinds of inferences you use as examples – since to examine is to apply general principles. General principles are inductive inferences. "The problem of induction" means there is no deduction to justify such principles.

You statement is self-contradictory.

My quotes were directly from the Stanford text and after reading them well. I was, of course, making a debating point, Julia sadly had only one thought method – tremendously inflexible and straightjacketed. But it shows a major problem arising from present educational didacticism. You learn this because this is how it is.

Julia: Elizabeth said: "I am not limiting my thinking to whether they perform their experiments logically, accurately, honestly or well, but

I am thinking of the role which science plays in society and how we can evaluate or quantify the qualitative aspect and impact"

I already pointed out that is not 'good science' it's 'what good does science do?" – your statements are invalid if it is about 'good science' – Good science is about the logic of method – if you do not confront the 'problem of induction' your talking nonsense.

More utter inflexibility. Here she wants "Good" to have a specific definition pertaining to its following a prescribed thought process. So the rubbish in, rubbish out is totally OK to her, so long as it's followed this logical system. Dangerous indeed, as I've shown elsewhere.

I then answered Liz' contribution which I've lost, but the gist is there:

Me: Thanks, that's cool, but I call that Bacon bit "reasoning"!

The morals of science are another place for evolution. My dad (again!) was a research zoologist/immunologist and so I grew up with visits to "The Animal House", where rabbits, rats and mice were kept prior to and during experimentation. Example to kids has unintended consequences and, in me, there was no wish to copy, as it all seemed sadly inappropriate. So my science had to be different, holistic and systems oriented not brutally reductionist.

So your last paragraph is as with my unease, too. From Einstein came the atomic bomb and, indirectly, Chernobyl and Fukoshima and there are so many similar, scenarios but, now, commerce drives, for example GM seed, governed by little other than seeking monopoly capitalistic dominance. Simple moral considerations are so squashed.
Then the horrible question – do good scientists work for Monsanto?

Elizabeth: Chris I was saddened to learn from a bright graduate friend that when he began to look for work he found that all the 'good' science graduates had only the opportunity to work for the 'big guys'.

Since he could not accept the (im)moral aspect of the work he was offered he has joined the (growing) ranks of the unemployed graduates.

Me: Yes, that's where we are. There is a fantastic amount of regenerative, systems sustainability work crying out to be undertaken but cold economics will not allow it. To me it seems, however corny it sounds, that we're on the cusp of a new age of enlightenment, Renaissance style. Certainly that's whence my energies are channelled.

Tell your friend to hang on in there, eh?

Pleased I kept this – it's exactly as I titled it but shows a certain upbringing and background and, if it helps me understand the GMC "Evidence based medicine" system, then so much the better.

5 – On Vitamin C

29th November, 2013.

This from the Linus Pauling institute:

 "Vitamin C (L-ascorbic acid) is available in many forms, but there is little scientific evidence that any

one form is better absorbed or more effective than another. Most experimental and clinical research uses ascorbic acid or its sodium salt, called sodium ascorbate. Natural and synthetic L-ascorbic acid are chemically identical and there are no known differences in their biological activities or bioavailabilities".

Note, however, they refer always to L-ascorbic acid. This being one of the two stereo-isomeric forms, the other being D-ascorbic acid, which is of less use. Wikisay:

"Chemically, there exists a D-ascorbic acid which does not occur in nature. It may be synthesized artificially. It has identical antioxidant properties to L-ascorbic acid, yet has far less vitamin C activity (although not quite zero).

"This fact is taken as evidence that the antioxidant properties of ascorbic acid are only a small part of its effective vitamin activity. Specifically, L-ascorbate is known to participate in many specific enzyme reactions which require the correct epimer (L-ascorbate and not D-ascorbate)." which seems pretty much on the nail.

Commercial synthesis of ascorbic acid seems to be always as Vitamin C, using a mix of biological and chemical processes which ensures production of only the L form:

http://en.wikipedia.org/wiki/Reichstein_process

http://www.dsm.com/.../vitamin-c/industrial-production.html

Of pure fact, rather than his more famous fiction, Isaac Asimov wrote:

"As it happened, Vitamin C was finally isolated by someone who was not particularly looking for it. In 1928, the Hungarian-born biochemist Albert Szent-Gyorgi, then working in London in Hopkins' laboratory and interested mainly in finding out how tissues made use of oxygen, isolated from cabbages a substance which helped transfer hydrogen atoms from one compound to another.

"Shortly afterwards Charles Glen King and his co-workers at the University of Pittsburgh, who were looking for vitamin C, prepared some of the substance from cabbages and found that it was strongly protective against scurvy. Furthermore, they found it identical with crystals they had obtained from lemon juice. King determined its structure in 1933, and it turned out to be a sugar molecule of six carbons, belonging to the L-series instead of the D-series. It was named "ascorbic acid" (from Greek words meaning "no scurvy")."

Asimov I. 1972. Asimov's Guide to Science. New York. p. 690-**700 Basic** Books Inc.

And Wikisay, re body concentrations and the physiology:

With regular intake the absorption rate varies between 70 to 95%. However, the degree of absorption decreases as intake increases. At high intake (1.25 g), fractional human absorption of ascorbic acid may be as low as 33%; at low intake (<200 mg) the absorption rate can reach up to 98%.

"Ascorbate concentrations over renal re-absorption threshold pass freely into the urine and are excreted. At high dietary doses (corresponding to several hundred mg/day in humans) ascorbate is accumulated in the body until the plasma levels reach the renal resorption threshold, which is about 1.5 mg/dL in men and 1.3 mg/dL in women. Concentrations in the plasma larger than this value (thought to represent body saturation) are rapidly excreted in the urine with a half-life of about 30 minutes. Concentrations less than this threshold amount are actively retained by the kidneys, and the excretion half-life for the remainder of the vitamin C store in the body thus increases greatly, with the half-life lengthening as the body stores are depleted. This half-life rises until it is as long as 83 days by the onset of the first symptoms of scurvy.

"Although the body's maximal store of vitamin C is largely determined by the renal threshold for blood, there are many tissues that maintain vitamin C concentrations far higher than in blood. Biological tissues that accumulate over 100 times the level in blood plasma of vitamin C are the adrenal glands, pituitary, thymus, corpus luteum, and retina. Those with 10 to 50 times the concentration present in blood plasma include the brain, spleen, lung, testicle, lymph nodes, liver, thyroid, small intestinal mucosa, leukocytes, pancreas, kidney, and salivary glands.

"Ascorbic acid can be oxidized (broken down) in the human body by the enzyme L-ascorbate oxidase. Ascorbate that is not directly excreted in the urine as a result of body saturation or destroyed in other body metabolism is oxidized by this enzyme and removed."

And I've not said a word about therapeutic use, for anything from scurvy prevention up to the orthomolecular megadose applications. These are there, of course, but this was background and answer to those who use Vitamin C "wholly as biology makes it" extractions. For such a simple chemical I feel this has no validity, particularly as the Dextro form is not co-synthetised with the Laevo rotatory isomer and so the latter as sold is pure and thus truly is "nature-identical".

6 - Putting Mars through some Wiking – another Saga.

May 17, 2012

"Food energy – supply and actual uptake."

This is framed as a chat with a Wikier – online encyclopaedia composer - but is, in truth, my still clarifying to myself some baseline elements of human sustenance.

I grew up, for complex reasons I'll cover here soon, very aware of the inadequacies of the digestive system but have never really come back to defining the absolutes. I've always also felt there to be too little clear understanding of this subject. As I've mentioned before, the late Oxford Prof Sir Peter Medawar talked of the desperate need to draw ideas together, draw disciplines together. This is a cross curricular exploration and I think it shows!

Well, in such an investigation the first stop is so easily taken. To research the energy contents of food, I went to Wiki:

http://en.wikipedia.org/wiki/Food_energy

which said:

"the convention is to use the heat of the oxidation reaction", that is the chemistry in the bomb calorimeter.

"This method of estimating the food energy has several defects, the most serious of which is that protein is not oxidized in the body as in the bomb calorimeter, with the possible exception of severe starvation. In normal conditions, the protein is metabolized in processes which require energy such as protein synthesis or replacement, synthesis of hormones, nucleic acids, etc. Thus, the food energy derived from proteins could be zero, if the energy saved by the body in using the proteic food components, instead of synthesizing them, is taken into account."

However:

"Each food item has a specific metabolisable energy intake (MEI). This value can be approximated by multiplying the total amount of energy associated with a food item by 85%, which is the typical amount of energy actually obtained by a human after respiration has been completed"

Then, though, from this high point, reality creeps in:

"In general, the efficiency of muscles is rather low: only 18 to 26% of the energy available from respiration is converted into mechanical energy. This low efficiency is the result of about 40% efficiency of generating ATP [the body's basic energy packet, adenosine tri=phosphate] from food energy, losses in converting energy from ATP into mechanical work inside the muscle, and mechanical losses inside the body. The latter two losses are dependent on the type of exercise and the type of muscle fibres being used (fast-twitch or slow-twitch). For an overall efficiency of 20%, one watt of mechanical power is equivalent to 4.3 kcal (18 kJ) per hour."

Actually I think you mean "at" an overall efficiency of 20%. Look:

http://en.wikipedia.org/wiki/Watt#Confusion_of_watts.2C_watthours.2C_and_watts_per_hour

"Confusion of watts, watt-hours and watts per hour.

"The terms power and energy are frequently confused. Power is the rate at which energy is generated or consumed.

"For example, when a light bulb with a power rating of 100W is turned on for one hour, the energy used is 100 watt-hours , 0.1

kilowatt-hour, or 360 kJ. This same amount of energy would light a 40-watt bulb for 2.5 hours, or a 50-watt bulb for 2 hours."

Remember – "One Watt is the use of one Joule per second"? This is that!

And examples:

"A person having a mass of 100 kilograms who climbs a 3 meter high ladder in 5 seconds is doing work at a rate of about 600 watts. Mass times acceleration due to gravity times height divided by the time it takes to lift the object to the given height gives the rate of doing work or power.

"A labourer, over the course of an 8-hour day, can sustain an average output of about 75 watts; higher power levels can be achieved for short intervals and by athletes."

Are you starting to see things more clearly, my little light bulb? Over the 8 hour working day this is 2160kJ of energy usefully converted (60x60x8x75). At 20% efficiency that requires consumption of 5x2160kJ = 10800kJ (aka 2580Cal/kcal) plus enough to cover base metabolic rate.

And we have also to look at efficiency of absorption from the intestine and of metabolic pathways and availability of nutrients held within the food.

Not quite like a big powerful car:

"A medium-sized passenger automobile engine is rated at 50 to 150 kilowatts – while cruising it will typically yield [provide!] half that amount."

OK, so you consume your Mars bar with its high energy content but, although it contains enough energy to lift you to the top of a mountain, in practice you're less efficient than a 1930s coal fired power station. Mind you, your pollution is nowhere near as bad and you do have an inbuilt combined heat and power system.

But your metabolism has a maximum power output, sustained by the conversion to energy of elements within the food that you have eaten. So:

"For example, a manufacturer of rowing equipment shows calories released from 'burning' food as four times the actual mechanical work, plus 300 kcal (1,300 kJ) per hour."

OK, I go with that, more or less – except where do they get 1300kJ figure from? But then:

" which amounts to about 20% efficiency at 250 watts of mechanical output."

Which is anything but clear. Sadly it then says:

"It can take up to 20 hours of little physical output (e.g. walking) [What cheek!] to "burn off" 4,000 kcal (17,000 kJ) more than a body would otherwise consume." This is questionable for using the 1300kJ/hr background cited above one gets 13 hours – without any walking at all!

Let me tell you that in twenty hours of walking I will have covered well over one hundred kilometres and used up one considerable quantity of energy. As well as an amount equivalating to my base metabolic rate. But what a bizarre statement anyway, and what has it to do with anything else in its proximity?

Then I think "Ah, bless the little volunteers for they know not what they say". And then I think about the abstract nature of information, fact and encyclopeadiae and that no fact is true fact until it is related to others at which point it may just vanish.

Before it does, though, I summarise:

The Mars bar, quoted post calorimeter, gives up 1500kJ/100g. Yum. However you might only harvest 85% of this. Leaving 1275kJ/100g. And then it goes through your metabolic pathways, your glycolysis, your Kreb's Cycle and , in your blood stream, travels as ATP type parcels to the point of use where the muscles are contracted, heat generated, neural signals released and so on. In all this, another 80% of the energy goes up in non productive, support services.

Leaving just 255kJ of muscle action. At steady, hard working, adult male rate of 75W that'll power you through 57 minutes. Thus, and assuming it weighs 100g, "A Mars a day keeps you going for less than an hour!"

7 - Weinstein's Question

November 5, 2012

"No homeopathy, just care, attention, good nutrition and absolutely NO JABS",

I answered to an online questionnaire, set on parents' attitudes to quantity which, if any, vaccinations they'd consider for their kids, by Daniel Weinstein, who maybe regretted this step later on!

Daniel Weinstein: Care, attention and good nutrition are essential for everyone. However, viruses and bacteria don't really monitor your habits. Vaccination works. It saves lives. It's silly to reject them.

Chris Hemmings: The point, my friend, is that a healthy immune system in a healthy body can do all the monitoring it needs to and establish recognition of countless potential agents of risk – be they bacterial, viral or other micro-organism. Vaccination is a crude, blunderbuss approach, never once having been subjected to a double blind or other controlled trial.

Look around at, say, the US population today. Picture of health? 25%adults diabetic and that figure in exponential increase. Obesity rife. Asthma, autism, allergies are a plague of your children and young adults.

Great times for those with shares in Pharmaceutical companies, mind you.

Daniel Weinstein: Chris Hemmings Ah, the Big Pharma conspiracy. I should have known.

Chris Hemmings: Wow, something you can really get your teeth into. Do I use the C word?? Of course not – there is no "conspiracy".

Whyever should they bother to conspire – they've got an ongoing market, for heaven's sake, so they'll supply it. That's business common sense as you well know.

But by shouting about such a scarlet herring you carefully avoid having to address my substantive point – namely the dire state of health demonstrated by the population of the US of A.

Which gets worse every day.

Andrea Leong: Some people are unhealthy due to lifestyle, therefore we all shouldn't get vaccinations? Chris, you are the one who needs to explain more fully. I'm pretty sure some healthy people contract vaccine preventable diseases. Vaccines are one of the easiest things we can do to lessen the incidence and severity of contagious disease.

Chris Hemmings: In 1980 5% of Americans were diabetic, in 2009 over 25% were diabetic and the curve has gone exponential. That is not "some" people, Andrea, that's one in four and rising rapidly.

If I governed such an abject disaster I'd panic. And then I'd look at all the probable and possible causes of such a gross deterioration in health standards over a period of unrivalled prosperity.

And yes healthy, unvaccinated people do contract these "vaccine preventable" illnesses. And they recover quickly and completely leaving within the body a lifelong lasting memory of the incident. This is the immune system, developed over four billion years and compromised by meddling in just 200.

Vaccines, it seems to me, are the simplest way we have to compromise the medium term health of the population and do away with its long term health totally. We could so easily work, instead, to boost innate, natural responses by ensuring optimal nutritional standards, good living conditions and informed nursing and medical care, to assist people through periods of infection.

Or would you wait until the whole population is diabetic, obese, asthmatic, allergic etc, etc……?

Andrea Leong: Argh, this page! Responses did eventually show up for me, but after I typed a reply, the page refreshed and cleared it. I'm on fb groups Stop the AVN and Informed Citizens Against Vaccination Misinformation if you want to continue the discussion.

Daniel Weinstein: And Chris, I don't really want to use this page to discuss the health of the nation with you. You believe in the naturalist fallacy. I get it. Be natural and all will be well. Like it used to be in the good old days.

Chris Hemmings: No, Daniel, I look forward, not backward. We live in an era where we have a vast capacity to master so many problems and yet modern medicine has as a key foundation an untested, profoundly damaging quack remedy poorly copied from peasant behaviour by a 19th century fraudster.

Surely the health of the nation is what this whole issue is about but I'm happy to agree to differ so long as you don't misunderstand, mislabel and misquote my stance. Vaccination is a global disaster area and the sooner that's faced up to the better for all of us.

Daniel Weinstein: Chris Hemmings Do you know anyone who has had polio?

Chris Hemmings: Of course. But you keep changing the focus and keep looking backwards. Enjoy.

Lara Lohne: Chris Hemmings, you say, "Look around at, say, the US population today. Picture of health? 25%adults diabetic and that figure in exponential increase. Obesity rife. Asthma, autism, allergies are a plague of your children and young adults."

However, your assumption that these diseases and disorders are somehow linked or the result of vaccination is flawed. Let me point out where the flaws lie for you:

1.   Diabetes: there are two types of diabetes, type 1 or junile diabetes, and type II or what used to be called adult onset diabetes. They no longer refer to it as adult onset because more children are developing it now then before, so the type II isn't strinctly limited to adults. Type I diabetes is typically present from birth, even though it can take several years to find it. Type II diabetes is the result of lifestyle choices, ill healthy and poor eating habits. Nothing more then that, and no vaccine will trigger or hinder the development of diabetes.

2.   In many cases asthma is also due to lifestyle, as it is more frequent in children exposed to second hand smoke. It can also be genetic. I had three children who were prone to night time asthma attacks, typically when seasons and weather changed. Their dad was the same way when he was little. They grew out of it and none of them have any problems with asthma. I on the other hand has asthma like symtpoms, but it is not technical asthma because it is due to damage my lungs suffered during a bout of pertussis when I was 17.

3.   I wouldn't say allergies are on the rise either, but that we are more able to test for them and better able to provide releif from them also. Allergies are just a mild immune response to something that our system is sensetive to, I think if we actually looked back at history, we'd find that many people suffered from allergies of one type or another, but it wasn't ever considered anything to be concerned about. Of course things change, and I think a lot of it is

over dramatizing things and making them out to be worse then they really are, which in and of itself is unhealthy.

4.   You likening autism and these other diseases and conditions to deadly disease (the plague) is, to be blunt, ridiculous and insulting. There is nothing deadly about autism and it isn't anything to be afraid of either. Multiple studied done for more then a decade have found there is no causal link between vaccinations and autism. In fact research is showing more and more genetics behind it and there isn't the exponential increase in autism like many would like us to believe. Studied done both in the UK and in the US have found, many adults are being identified now as having an ASD, but as children they were severely learning disabled or mentally retarded or in cases of high functionality, completely overlooked. The diagnostic criteria in place now is more indepth and detailed to better identify the multitude of traints that are present in any disorder found on thr autism spectrum. When comparing numbers of adults who have been identified as having an ASD to children being diagnosed now the numbers are remarkably close. Autism is a neurological disorder of development. The brain is constructed differently, therefore wiring is different. No amount of biomedical treatment will do anything to change the contruction and rewiring of the brain. Proper therapies can however. And please alsp try to keep in mind, autism sepctrum disorders are developmental delay, not stasis. Even without interventions a person will progress. With proper therapies they will progress significantly faster. My youngest child has autism. He was born with it, showed differences from my other five children from birth. It wasn't until he reached a developmental platue and began to regrss in some skills that we realized why he was different, but we knew he was different from birth.

What it all comes down to, with the examples you've given is poor lifestyle choices or better medical diagnostic methods. It has nothing to do with vaccines.

Chris Hemmings: "A" plague, not "the Plague", Lara.

But this site has twice wiped off my reply to you so ciao. [*This was indeed most frustrating. Later I found the conversation – one of many, many which developed out of Daniel's original question had switched itself to my home page – thenceforward matters were straightforward.*]

Peter van Loon: Chris Hemmings "Vaccination [...] never once having been subjected to a double blind or other controlled trial." *cough* bullsh*t *cough*

Chris Hemmings: OK, Peter, you try to find any but I tell you there have been no such trials ever done on the effectiveness of these concoctions. Apart from anything else, it's impossible!

Peter van Loon: I thought you were talking about safety, not effectiveness. But reading your post again, it's so rambling that it's impossible to tell. So I don't know whether you're moving the goalposts or just bad at expressing yourself.

Chris Hemmings: Wriggle, wriggle.

Chris Hemmings: What's unclear about: " Vaccination is a crude, blunderbuss approach, never once having been subjected to a double blind or other controlled trial." ?

Andrea Leong: It's unclear whether you mean trials for safety or for effectiveness (in which case I would assume you meant both), but when Peter asserted that such trials had been carried out, you replied that they hadn't been done for effectiveness. Thus it appears you've already accepted the safety of vaccines, yet this is at odds with your description: "untested, profoundly damaging quack remedy poorly copied from peasant behaviour by a 19th century fraudster".

But this is all moot, since it is clear from your overall body of posts that you believe vaccines are neither safe nor effective.

Btw, this site seems to be working fine for me know. Pity, because I don't like what I see.

Chris Hemmings: Ah, but we don't always like what's good for us do we, Andrea?

BTW, there's no way I just said "they hadn't been done for effectiveness". That was Peter, bless him, reading double Dutch in some rambling manner. My statement is crystal clear. And true.

Andrea Leong: You're right, Chris, I don't like getting needles, but I got my whooping cough booster a few months. Guess what? No ill effects apart from a warm shoulder for the rest of the afternoon (although I suppose someone will tell me that my health has been forever compromised).

Chris Hemmings: Hopefully not – but was there any benefit?

Anyway, if you're "becoming a scientist" that very phrase is deeply in your favour so, yeah, just keep an open mind.

I guess we should stop "needling" each other, now – enjoy your studies.

Lara Lohne: Chris Hemmings either way you spell it, a plague or the Plague makes little difference. It still infers deadly disease and suffering and that is not autism, nor it is any of the other disorders you find so offensive that people have.

Jamie Masciantonio: and Type II or adult onset diabetes is not necessarily poor health choices. Genetics still plays a roll in that, as well as other health problems (such as a high number of people with thyroid disorders are also becoming diabetic later in life)

Chris Hemmings: I refer you, Jamie, to a graph I posted a few weeks ago: [This, from CDC shows the alarming graph, the shocking exponential curve in diabetes cases in the USA on page three: https://www.cdc.gov/diabetes/statistics/slides/long_term_trends.pdf

Chris Hemmings: Now that graph has entered exponential phase in just thirty years. This cannot in any shape or form be taken as indicating genetic change. You'd need hundreds of years and a vast selective ADVANTAGE from diabetes for that to occur!

Chris Hemmings: And note the originators of the information shown are the United States Centres for Disease Control and Prevention.

Chris Hemmings: Lara, plague is "a pestilance, affliction or calamity on a large scale".

Chris Hemmings: But, anyway, I do not find all these disorders "offensive" – in fact I have profound sympathy for the sufferers and for their families. And, indeed for society at large as these conditions can clearly all share a common causative factor – the modern dash for supposed preventative vaccines rather than traditional nursing care.

Erwin Alber: I'd rather be called silly than having to live with a vaccine injury, or ending up with a vaccine-injured child, Daniel.

Lara Lohne: Autism is not a disease. Therefore for you to lump autism in with things like obesity, diabetes, etc. is naive and ill informed on your part.

Chris Hemmings: Lara I used the word "disorder" which covers the very wide autistic spectrum, from Asperger's right down to the more catatonic and destructive behaviours. Disorder also can be used for diabetes and obesity. However the term "disease" would be equally applicable for all three.

Whilst I remain civil with posters strongly supportive of the vaccination philosophy, it seems that you are not willing return the same courtesy. I am neither naive nor ill informed in these matters, as is quite clear from the answers I have already given. In fact I have carried out twenty years scientific research and am very well and widely informed.

I doubt, however, that you have anything like as much experience, information or understanding of the issues involved. Clearly not, in fact, from the nature of your postings.

Lara Lohne: There is a big difference between the terms disorder and disease. Disease implies some amount of control in whether or not you get it. Disorder is otherwise. That being the case, Obesity is a matter of choice because it can be completely avoided by making wise and healthy lifestyle choices. The same isn't always true of diabetes as type I is typically from birth, and genetics can play a significant role in the development of type II. However, there are lifestyle choices that can be made to significantly minimize your genetic risk to type II diabetes, therefore, also not a disorder. Autism, on the other hand, there is no control over whether or not a person has it, just as their isn't any control over whether or not a person will be born with Down Syndrome. That is the difference between a disorder and a disease; control over the outcome at least to a degree. Autism is genetic, and until they perfect a prenatal test for it, like they have for Down Syndrome, the number of children born with autism will not change. Having a child with autism myself though, I don't feel a genetic disorder is any reason to terminate a pregnancy. I know others will disagree with that. I have learned so much from my son that I never would have learned if he hadn't been born the way he is. I love him for who he is, and while he has autism, autism doesn't have him, and he is more then his diagnosis. Therefore, your ignorant statements that autism is a disease are highly insulting to every autistic individual ever, and there have been a lot then you are even aware of I'm sure.

Chris Hemmings: Autism is NOT a genetic state. If it were, how come there was no autism prior to around 1945? Sudden beneficial mutation all over the World? And yes, I am a geneticist.

Down's syndrome is, of course, a duplication of a whole chromosome and so clearly a genetic disease or a chromosomal disorder. They're only words and they are what we use to describe the world and communicate with each other.

A reference on "Genetic diseases" for you :
http://www.medicinenet.com/genetic_disease/article.htm

As for diabetes just look at graph I posted a bit earlier – 8 posts up – and see a 500% increase in thirty years. There's NO genetics in that but we urgently need to isolate all the lifestyle choices and changes that have caused it. I'm confident that I can untangle the threads involved.

Anyway, whilst I have profound sympathy for your personal circumstances and I admire the way you obviously work positively and constructively with your child, your continued use of gross and coarse insults does not enamour me of you. Nor does it support any case that you might construct. Be civil or shut up.

Genetic Disease Causes, Types, and Conditions:
http://www.medicinenet.com/genetic_disease/article.htm

Martin Bouckaert: I'm sorry, Chris, but my autism is most certainly not a disease. 20 years of scientific research? 20 million years of researching the wrong thing and getting the wrong information isn't worth two minutes learning about it from an accredited professional. If you think autism can be categorised as a disease, then you are, indeed, misinformed. Autism is not a disease, there is no epidemic or plague of autism, and no one suffers from autism. The only thing autistic people suffer from is morons that fail to understand it.

Chris Hemmings: Martin, I really have no concern as to what you name "the condition" or, even, the nest or spectrum of conditions. Please, you suggest something – I find it hard to relate to just a negative.

As for your faith in "accredited professionals" the whole point of research is that you look at a range of viewpoints, of data, of case studies and case histories. You do not just go to a single agency or individual.

Martin Bouckaert: It's called autism. It's not what I name it, it's what it's called. It's just like how a car is called a car. It's not called a car because someone wants to call it a car, and you don't see people running around calling it a gollywog or something they've just decided to call it on the spot. This isn't about what I want to name things, or how easy or hard you find it to relate to things (which I might add, if you are allowing how difficult it is for you to relate to influence your opinion, then why are you acting like it's not a biased one and it's a bottom-line fact?).

It's about who's done the work, and the science, and knows how to interpret the data correctly because they've received training to do so, and have the credentials to prove it. That's WHY credentials EXIST in the first place. It's not about faith, that's WHY we make SURE that people have credentials – if it was just about faith, then we'd be listening to just about anyone that's "done their research." I think if you are going to pretend like you, with your lack of credentials, stand above someone with credentials in matters of science and research, then that is simply arrogance, and I will tell you now that I will place more "faith" in someone with the credentials to back up their experience and work than someone who does not have the credentials.

It's absolutely not about credentials, it's about what you can prove that you are capable of, and about what you can demonstrate you know. You are no expert, and it shows in your dismissiveness of the experts. Do I detect a little jealousy? Competitiveness?

Chris Hemmings: Prior to around 1950 the term autistic was not used and there were few with the range of behavioural characteristics now associated with this descriptor. Now the "diagnosis" is common and many fall into this category. "Autism" is found in many differing situations and for many, many different states of being. You could equally say this about cars, I agree.

You write: "It's absolutely not about credentials, it's about what you can prove that you are capable of, and about what you can demonstrate you know."

I so agree with that and, in this topic I'm more and better "expert" than most.

8 - I didn't know this but........

February 25, 2013

A matter of life and death – and conviction?

The terms and tones of debate are useful, as they demonstrate the mindsets of the players. They also clearly show where faith is based upon realistic foundations. "Unvaccinated America" posted the following discovery on 5 August 2012:

I didn't know this until today! Did you? If a person tells you that the link between vaccines and autism has been scientifically disproven, kindly remind them that the man responsible for allegedly disproving the connection between the MMR vaccine and Autism has been indicted on fraud charges by the United States Department of Justice. He never did the study, and the data he submitted was fabricated. He bought a new house and a Harley motorcycle instead:

http://www.justice.gov/usao/gan/press/2011/04-13-11.html”

Susan Faia Eaton

That's why I never listen to any pro-vaxers. My children are not their Guinea pigs. My reason to not vaccinate had more reasons than just high chance of getting Autism but permanent brain damage, serious illness, cancer, allergies and low immune system thus getting sick all the time, then having to be on more drugs. Oh and death!

Kyle Lloyd Munson

Yeah but what they fail to tell you in this article is the timeline of events. The alleged "fraud" by Thorson did not happen until at least a year after the publication of the thimerosal study which was validated and still holds up. [Clearly not a true statement on several grounds eg it was not a thimerosol study talked of!]

Tammy Monroe McMullen

I still believe in vaccinating my kids. I also believe that the MMR shot DOES cause autism if given too close to other shots. You can't protect from everything and not everything is good for us I understand. But safety belts in a vehicle can also cause death...IN RARE CASES!!! That doesn't mean every seat belt will kill every person who usues them! It just means everyone who fails to use one could die faster than those who do use them. It's a percaution just like many other thing's.
Nothing is fail proof! Helmets while on bikes, pedal or motor, can HELP you in an accident...depending on the accident. So are you going to be the one NOT using a helmet? I wear mine every time! It's a choice. Each of us have to make the choice for ourselves unless a law is in place...then suck it up! Until then, we all are trying our best with the info provided. Good luck with your choices.

Chris Hemmings

Tammy, you distract by using an analogy which has poor synergy with the matter in hand. Vaccination is a direct intervention with the metabolism of the body with lasting physiological outcomes, using powerful (bio)chemicals, whereas a bike helmet you can just take off after your ride.

There is often legal as well as strong social and economic pressure being exerted by the vaccination lobby and the above case demonstrates just how far they are prepared to bend rules and lie to support their position.

Several other pro-MMR studies as published in the Lancet etc are also deeply flawed as they fail to approach the subject objectively. For example:

1: A large study which looked for reactions to the vaccination only noted those arising in the week following the date it was administered.

2: A comparison between children who received the MMR with other children who did not and finds no difference in numbers of autistic cases arising fails to point out that the second group had received a range of other childhood jabs which can and do precipitate the same autistic outcomes.

Genevieve L Ferrantino

Good to know

Sean Cannon

Would you rather have an autistic child or one that dies of Mumps or Measles?

Sarel Botha

Sean, your chances of getting those diseases are very small and a very small number of people actually die from those diseases. Also, do you know how many children die from vaccines? http://vactruth.com/2011/05/05/infant-mortality-rates-increasewith-vaccines/

Sarel Botha

There were many studies, but there are problems with all of them http://www.14studies.org/studies.html

The study the medical industry will never do is to compare vaccinated and non-vaccinated children http://www.vaccineinjury.info/vaccinations-in-general/healthunvaccinated-children/survey-results.html

Unvaccinated America

It is MUCH easier to recover from mumps or measles than to heal from autism. There is no comparison in my opinion.

Tammy Monroe McMullen

All of us have great points to our thoughts and opinions. However, I stand by my opinion. Yes helmets and seatbelts you can take off. So what?! The point is the same...NOTHING IS FAIL PROOF! Apparently, someone didn't read and comprehend ALL of my comment. Measles and mumps are and illness. Autism is just a change. I wouldn't want my children to have ANY of the mentioned or any other's for that matter. AGAIN I SAY IT'S A CHOICE FOR ALL OF US TO MAKE FOR OURSELVES.

Rus Oister Jr

The reason why there is a small chance of getting measles or mumps is because most people get vacated for them. If we stop giving vaccines we will see them more often. I guess none of you give your kids processed food or let them drink tap water, I am sure you all have breast fed your kids as well because who knows what is in the formula.

Sean Cannon

I got all my vaccines and so did my kids, and we're all happy and healthy. To each his own. Some infants probably react negatively to the vaccines, and some would likely die without them.

Chris Hemmings

Tammy, you cannot remove a vaccine and the damage stays with you for life. Mumps and Measles are just two weeks illness, autism and other vaccine damage don't go away – you remain "changed". If vaccines are compulsory then, to me, that's state inflicted and all for no benefit as our bodies' immune systems are already totally able to fight off these infections.

Rus, Sean and Jake? No, I'll leave them to do some research someday. Hopefully!

Tammy Monroe McMullen

Chris you are entitled to your opinion no matter who agrees or disagrees with you. I bet you got all of your vaccinations! Go ahead and NOT vacinate your kids. That's your choice. But to love your kids totally and want to protect them to the fullest...except every illness that could kill them! Doesn't make sense to me. Make sure to only wear clothes you made by hand, breast feed your children, make and grow your own food and home school your kids...I don't want your kid's around vaccinated kids including mine, so THEY become sick and die and you bitch saying we did that!! Stop bitching about my comments because I heard your opinion the first time! Find someone else to rag on!!! I STAND BY MY VACCINATED KIDS AND SELF!

Christina Kirtley

That is crazy!!

Tammy Monroe McMullen

Chris I understand you have been educated highly on this topic and probably many other's but I don't care. Chemotherapy is a poison used to kill a bigger poison...cancer! Same difference. We do the best we can with the info provided and found. My kid's are proudly vaccinated!!! I am proudly vaccinated!!!!! My dog is proudly vaccinated!!!!! Do you get it now? There isn't an arguement in the world you will win over me!

Andrew Moyer

My brother who is a pediatrician I might add says you are putting your kids more at risk not getting your kids vaccinated than the vaccinations themselves.

[And he then brings his brother into the chat:]

Alvin Moyer

This whole debate is occurring because of the success of these vaccines. If people remembered the thousands of children who died each year from vaccine preventable diseases, there wouldn't be a debate. And FYI, it is Paul Wakefield, the man who first published a "suggestion" that MMR vaccine and autism were linked, who has been indicted on fraud and had his medical license revoked. Thousands of studies have been published finding no link.

Jo Ann Steinmetz

Keep raising awareness about the truth, Alvin.

China Gushiken Omg!

Sound Mamas

Yes! Spread the truth!

Chris Hemmings

Alvin, whilst you are correct that the UK General Medical Council, GMC, took away ANDY Wakefield's doctor's certificate that is not the issue here.

Dr Poul Thorsen is the one indicted by a federal grand jury on charges of wire fraud and money laundering based on a scheme to steal the grant money that the Center for Disease Control, CDC, had awarded to governmental agencies in Denmark for autism research who's wholly flawed and compromised "disproval" of Wakefield et al's work was then used by the medical establishment against Andy.

Simply the evidence they site is based on flawed sampling and skewed comparisons which demonstrate nothing other than a way for fraudulent researchers to tap into bottomless pots of funding to sample at will.

Oh, and Tammy this is not a "who can shout loudest" competition and frankly I do not care whether I can provide simple arguments on this site to convince you to balance your views. It's others I care about, those with open minds and, most importantly, the generations not even born yet. It's them I'm mostly concerned about as I want them to be born.

Andrea Murch Stevens

Oh my gosh! I am laughing at the ignorance on this thread. Obviously the pro vaccinators haven't done any research since they are just spouting off pro vaccine propaganda.

Carolyn Bursle

Thorsen's study is not the only one disproving a link. So there's a fraudster on each side, but the real science is overwhelmingly on the vaccine side.

Jodie Tysver Jensen

A different perspective? Consider the source. It tends to be hurting, regretful parents of damaged children who voice their strong opinion that vaccines are potentially very dangerous. It also tends to be drug companies, with something ($$) to gain, that insist on the harmlessness of injecting chemicals into tiny bodies. Parents have nothing to gain except sharing a heartfelt story to help others avoid such tragic consequences. When you side with anti-vaxers, they reap no benefit. When you side with pro-vaxers, the difference is a larger bank account for them...

Which is where the conversation ended. Part of me wants to add a commentary to highlight matters I see it illustrating and how opposing camps are structured. For now, though, I think it's illustrative enough as a stand alone entity but away from the initial high speed and heat of the exchange.

9 - So? Annul the Institutionalised Bias!

March 26, 2013

I feel like a DJ sampling sounds to play at a nightclub only here it's voices sampled from the ether that is the internet. An interesting discussion arose out of the autism and statistics piece I quoted as "The Null Hypothesis", recently. It developed as follows:

Me

Hi. As you say much of the debate between "the two sides" has been informed by epidemiological studies. Only "informed" is a poor descriptor to use – blinded might be better.

To disprove a connection between the MMR and development of autism a vast study looked at kids who had received the MMR and compared them to kids who had not been given that jab. They found no significant difference in numbers of cases arising in each group. Therefore, said the epidemiologists, and the medical establishment and the media and the politicians, the MMR does not cause autism.

How stupid do they have to be to believe that?

Look at 1940s kids – no autism. Look at never vaccinated kids – no autism.

And the animal studies have all been done – on a child, by child, by child, basis. That much is quite clear, too.

A scientist

Hi, Chris, thanks very much for your input. One thing I'd really like to stress is that it's vitally important that controlled studies be performed investigating not only whether there is a true relationship between vaccinations and autism risk but, if there is, what are the specific underlying biological mechanisms through which this occurs. But in order to answer either of these things, we need to have controlled lab-based research.

That, unfortunately, will not be answered by looking at larger trends in human populations (though it'd be much easier if this were true!). As part of the point of the above post is to stress that such large, vague studies don't offer us the specificity required to answer a disease-exposure relationship question; those answers it does give us are dubious and require more in-depth study.

I realize that non-scientists may have difficulty picturing why scientists can seem so anally-retentive about this, but I can promise you it's from many, many years of combined experience investigating cause-effect relationships and seeing pitfall after statistical pitfall.

Looking at large population studies, even the kind you propose, are still untrustworthy– because there are so many variables that will inevitably pervert the results – which is why I'm calling for more labbased science. Animal studies, cell culture, etc. With this design, there is far more control over all the variables so that if there is an effect, we will

1. be more likely to catch it, and

1. have a comparably easier time discerning potential causal relationships.

   Again, the purpose of the post is to stress that we need to have better science, not just more of the same.

   Me

   Gosh, there's a lot of follow up here so I'll try to be brief.

   I'm university genetics/immunology background and have 3 never vaccinated kids – that decision made in 1993, way before the MMR/Wakefield saga.

   Aware of the might of the orthodoxy in pressing home "the need to be vaccinated" I've worked hard to develop as broad a picture of the sundry inputs to this debate.

I think lab animal studies can provide little information to clarify the picture – although our household, handed down dog is a severe case of vaccine damage. (Honestly – she practically died from it, and remains affected several years later.)

Powerful chemical cocktails introduced to the bloodstream of very immature human children obviously have a range of outcomes. A bit like a drone bomber, piloted by an agent 5000 miles away going after an insurgent in a busy market square.

We should start off with the onus of proof of functionality and nontoxicity placed upon the vaccinator. That case has never, ever been demonstrated.

Keep your thoughts refreshingly open,

Chris

PS – "Herd immunity" is another sacred cow that needs dispatching. Are you part of a herd?

Me, next morning

This just came to my inbox, as a contribution to the discussion. How wrong:

"Finally, I think that the context of *other* environmental, or genetic participation in autism should be considered; namely, that everything we see seems to be a low penetrant effect. *If* vaccines are playing a part (?), I would expect them to follow a similar profile, i.e., a small nudge as opposed to a massive force."

Vaccines are, in a very clear manner, the spanner in the works and not a minor contributor to an overall trend. That there is a wide spread of outcomes is to be sorted out – and obviously the input varies from case to case for a range of reasons (diet, age, maternal immune status, breast feeding, home and lifestyle, other stresses. Even a genetic contribution!)

Autism was non existent in the 1940s and has grown on an increasingly exponential scale since then – matching the increased usage of vaccines.

I'm not saying QED but I am saying vaccines are an enormous elephant in this room – as per my blog!

Chris

Later again, Me

Hi. This is an area I've covered before and feel understanding is openly or passively avoided by the scientific research community. I freely admit my role is as agitator because the inactivity frankly appals me;

So from May 2011 there's "If Wakefield were a bond trader…" http://greencentre.wordpress.com/2011/05/13/if-wakefield-were-abond-trader/ but also, in a push to clarify things even further I just put this together – so, inspired not a little by this chat I can now offer you "The Annul Hypothesis": http://bmeandothersciences.wordpress.com/2013/03/06/the-annulhypothesis/

A scientist

Yes, I had read that this morning. [When I'd first posted it.] Thank you for the additional vote of confidence.

Me

Yes it is there, indeed – but please say you see my point.

A scientist

I can certainly see cause for concern and reason for wanting more research. But I am pretty agnostic when it comes to tending one way or the other, though you've spent more time reviewing this issue than I have. From what I've read you seem to be leaning towards a vaccination-autism link. I can certainly understand your point and we both definitely agree on the need for more and better research.

Though, as I say, I still want to see that research first before I take a stand on whether I think there is a true relationship or not. Hopefully you likewise can understand my ambiguity as much as I can understand your passion.

Me

Sure I understand scientific objectivity. That is my stronghold, "Deep in my DNA" as one might blushingly say, but it is fundamental to all my thought. As you correctly point out, the statistical analyses quoted do not and could not disprove an association between MMR use and physiological compromise, such as autism.

My point is that they are not even designed to, for they are conceived to demonstrate that the MMR is "a safe vaccine" and not to explore the far more relevant question as to whether vaccines generically are harmful. Should I write it loud? Vaccines GENERICALLY are harmful. Mm, that feels better.

It's a conjuring trick, a diversion, sleight of hand and always ignores any attempt to explain the soaring rates of physiological abnormalities developed post vaccine administration to infants. It dates back right to the time of Jenner and continues today.

There's obvious humanist and personal reasons for my concern but one of the deepest reasons is pure scientific because I detest the misuse of this noble discipline. Deeply entrenched here, the cavalier misuse is monstrous and the worst is that the scientific community is so cowed it does not shout loud the obvious. Yeah, that's my passion.

A scientist

I do understand your point. I suspect that, given the monies tied up in vaccinations, plus potential dangers of people foregoing vaccines, there have been inherent biases which have lead to seeking confirmation for minds already made up. Then again, those are only my suspicions. But it is the primary reason I wouldn't trust the CDC or like organizations investigating this issue because I would be distrustful of that potential bias. With the epidemiological studies, we may well be seeing that.

Perhaps to some that may reek of governmental or big-business paranoia which has coloured our country's history, but when it comes to capitalism I haven't generally noticed such paranoia to be particularly unfounded.

Insurance companies allow people to die so they can save a buck (or billions of bucks); such companies also eagerly lobby our government in a bizarre yet legal means of bribery; and government in general seems more interested in their own personal lives as career politicians, living election by election, rather than in their responsibilities as civil servants.

I suspect, though, that the greatest roadblocks have not been due to deliberate deception but to bias. The problem of people, both sides, making up their minds before research was available. Admittedly, if I were a parent and my child displayed a rapid regression of skills immediately following vaccination, I'd probably be convinced myself. And actually it is those cases which I find most moving and which allow me to keep an open mind and not side necessarily with the majority of researchers and doctors.

Subtle regressions are difficult to be certain of, but from my understanding there have been enough cases of coinciding severe regression within the one or two weeks following vaccination which are difficult to argue with. At least I find them difficult to argue with. It's those cases which stick out in my mind and which I wish were studied more intensively. I think they could be especially informative. [Interestingly, of course, it's these kind of cases Andy Wakefield and friends were investigating – and we know what happened to them!]

<u>Me</u>

Oh, there's no " potential bias" in those statistics, they are deliberately obscurantist – they ask the wrong question because it does not challenge the actual outcome of the jab, just the outcome relative to other jabs.

OK – for regression read the history of the Guardian's Charlotte Moore. Three boys she's had. One and two are well chronicled – by her – regressive, stage by stage autistics. The third, last born, is not. Although she gave MMRs to the first two, the third remains unvaccinated and in normal good health. When I last read her she still insisted MMR had nothing to do with the autism, even though' the regressions were each just following a further jab. Hey, though, she got lots of copy from it…. What an odd career move.

Then go to Australian Dr Viera Scheibner – http://www.vierascheibner.org/ to read about the timing of infant reaction to vaccine challenge. Like clockwork! Very scientific work and clearly deeply worrying. I've met her and she's accurate and honest.

For any benefit to vaccines I guess you need do your own research but bear in mind how well evolved and sensitive the immune system is (4 billion years in the making) and how homeostasis and other physiological processes are non-linear and interlinked. To me, it seems we should not seek to hijack the system, as we do thro' vaccines, but to enable optimal sensitivity – nutrition, lifestyle etc and accept that illness managed – measles, chicken pox etc – are natural strengthening of the system so a positive rather than a negative.

Finally the insurance issue. As well as the General Medical Council finding it impossible to admit to making such a profound and ongoing mistake and so deeply losing face, the issue of reparations for vaccine damage would bankrupt the Government, the

Pharmaceutical Industry and the Medical practitioners. Just imagine the claims that could be made for life-long-care and damages. It would make today's banking debts look quite small.

This also will sound trite but, clearly, the best cure for autism is to stop creating autistics. To return to the state of play circa 1940 when there were none.

PS. The solution for the ongoing cases of vaccine damage is also there but that's another story. Another tea break, maybe!.....

A scientist

Thanks for the additional info, although I would hazard some caution in word choice with this statement, "This also will sound trite but clearly the best cure for autism is to stop creating autistics. To return to the state of play circa 1940 when there were none." It can really be taken the wrong way, even if written with the best of intentions.

Even though I am a scientist and truly do feel for the plight of families who are struggling just to survive day by day (which I would gladly make efforts to help), I am also very close to the online adult autistic community and can appreciate aspects of neurodiversity too. Many people do feel that severe cases of autism have little in common with higher-functioning people, but while there are obvious differences, I see them as differences partly of severity and not of kind.

Although, on the other hand, I believe there is a CONSIDERABLE amount of heterogeneity across the spectrum, which includes very heterogeneous aetiology of people with comparable levels of disability. So when it comes to autism and the behaviours, I'm a lumper, but when it comes to aetiology, I'm an excessive splitter– but in neither case do I find level of functioning to be a useful paradigm in understanding the science of the condition.

Sorry for the digression but I'm familiar with some of the criticisms that may follow my ties with camps like neurodiversity and just thought I'd address them before they're asked. I don't necessarily consider myself an ND proponent, though at one point I could've probably been described as such. But I've had the opportunity to traverse a number of different camps and find myself more prone to understanding the passion and positives of each one.

I see the value in not perpetually treating another human being as a disorder and attempting to normalize their behaviours solely for the sake of society, but then I see no point in failing to help improve quality of life for people who may be suffering.

Me

Don't get me wrong, we cannot wind the clock back – I know that and would in no way propose it. But here my point is that in that era, pre nearly all jabs, autism was not known. That situation could be established today by removal of the causative factor – vaccines.

A PC aspect I have not toyed with but it's a shaky ground to ignore the issue. "Let's continue this practice because a low percentage of the recipients obtain enhanced, if rather tunnel vision, intelligence. The majority, of course, will continue to suffer lifelong physiological damage". I know some such taking PhDs, and they work with both dedication and application. But this proposal is an unnecessary Russian Roulette – for these intellects not impacted by artificial stimulus (vaccines) will still flower and maybe with greater strength, and also without any of the rest of the range of vaccine associated ailments, as well.

Always, the picture broadens but without that vision you cannot address the issue. In fact I rather feel that without the broader vision one's impact can be negative and the problem compounded.

I thought the conversation over but then:

A second scientist (Natasha)

Sullivan suggests: "We are now left with what I am admittedly simplifying to: "there is an immune component to some or much of autism. Vaccines affect the immune system. Therefore vaccines cause autism".

Says who?? The facts are that there is an immune component to autism, and that vaccines affect the immune system, but of course those facts do not prove that vaccines cause autism. What they do prove is that we are in serious need of some serious biological science here. And as PD remarked, knowing what goes wrong and why on biological level will point to things that will help kids in the future.

Me

Oops, me again. With Natasha I cannot but agree. The statement:

"there is an immune component to some or much of autism. Vaccines affect the immune system. Therefore vaccines cause autism" is flagrantly absurd.

However, as I hope I have clearly demonstrated, the clearest evidence is already there in abundance to demonstrate the profound impacts of vaccines on infant physiologies and, possibly less overwhelmingly, on adults as well. We cannot now waste time in creating in vitro laboratory models to describe these complicated interactions, which can have at best marginal relevance to this globally widespread problem. What is required is honest, objective assessment of the evidence we have and then a rebuilding of the healthcare system to accommodate this clear reality.

Which was anther end point.

It brought back to mind memories of my discussions with Professor Paul Shattock, a pharmacist who developed an interest in autism after his son developed the condition early in his life. Paul has worked in intensive and dedicated manner on the subject and has run a centre for research in Sunderland. We talked one day of the impacts of vaccine use and he was both torn and, so, equivocal. At my insistence that autistic outcomes were resultant from the whole process of vaccination he kept responding that we could only investigate detrimental outcomes on a jab by jab basis.

So, in looking at the MMR's impact, one had to ignore all other vaccinations. It was beyond the pale to do otherwise. Although he clearly had strong sympathy for my position, his implication was that to take my position courted ridicule. I pinched myself then and demurred, now, some five years later, I most vehemently decry that attitude as preposterous and a direct and institutionalised obfuscation!

But, as the above discussion demonstrates, even the most open minds cannot work in the absence of objectively presented data. The better will smell a rat, maybe, but still continue to follow prescribed direction, for that is as their roles are cast. I just spent a couple of hours drafting a précis of the chats but, really, I prefer to present the discussion in full. Just as the impact of vaccines on each recipient is different, for good, physiological and historical reasons, so each individual approaches the subject with a different set of preexperience and information. Objectivity is a path we have to walk, and many just don't.

Ho, hum, in jumped the Scientist, once more:

Anecdotal evidence is vitally important for informing the design of research studies in a case like this, but there are good reasons that decisions to drastically alter healthcare are based off of wellcontrolled science (ideally) and not anecdote. It's also vitally important that such changes be carried out cautiously because, even though many people now would not remember it, inoculations do actually prevent other horrific diseases, some of them far worse than autism.

Ironically in fact, several rubella outbreaks in the 1960s– the very virus for which part of the MMR is vaccinating against– were closely linked with increases in autism, mental retardation, schizophrenia, not to mention the usual slew of conditions like blindness.

I am not advocating sudden change in healthcare protocol, which I hope the above article made clear enough, though there are probably some precautionary steps we can take like reducing the heavy vaccination load infants are given or even waiting to vaccinate until after 2 years. I am indeed advocating for the kind of research, Chris you feel we don't have time for.

I understand the need for hurry, however I've also been around enough to know that, even though people may feel absolutely certain they know precisely what is occurring, they may well be wrong. And it's that science that you are so fervent we don't have time for which would help to clarify that.

This is not an easy situation, and to go back to the days preinoculation when a person could expect that about half their children would die in the first few years of life due to one contagion or another... well, I'm not eager to head back in that direction either.

We talk about doing things "naturally" but humans have only been living in HUGE communities, e.g., cities, for the last couple thousand years. This is how contagions are bred, through close contact. So one could also argue that living in such close confines with one another is not "natural"— not to mention the means we have now of travel and spreading those diseases.

But it's our current lot, so we must use the big brains we've been given to think up creative ways to reduce illness and fatalities which inevitably rise because of our way of life. One way is vaccination. But we need to learn our vaccine science better to figure out more accurately what are the risks versus the benefits.

So, Me, again!

Right. Now I do not propose retreat but progression. That's what humans do.

There is not "anecdotal evidence" to inform the design of research studies. There is a vast mass of evidence built up over 200 years but oh so accelerated in the last forty which is dismissed by the MedicoIndustrial Complex, MIC , out of hand. And yet you pick out an ironic anecdotal tale to point to the "dangers of German measles". Just the same scare tactics the MIC themselves use. I would suggest in that case that there could have been a pre-potentiation by prior vaccines which transformed what is normally a very mild illness into one so damaging. There is a long record of such instances, and I reckon SSPE is another.

The twentieth century saw, for Europe, America and other more affluent areas, a steady and profound improvement in living standards – housing, cleanliness, diet etc. Concomitant with this, death rates of children from childhood illnesses fell dramatically. All this prior to vaccinations. [Smallpox , Jenner and subsequent government vaccination campaign disasters you can read up on. They form a very sorry saga. See Leon Chaitow's "Vaccination and Immunisation".] At the time vaccines were introduced, death from these illnesses in such countries was rare and even those rates were still falling. I had measles, mumps, chicken pox and possibly whooping cough as a kid and loved time off school and lots of care and attention.

And so I developed strong, yes natural immunity for life and, yes, autism is for life too. Two weeks coddling versus a life sentence. Now which would one choose?

Anyway I'm not doing anything other than react to drastic changes in the "health care" protocol. All the current jabs have been foisted upon us with no evidence as to their efficacy or as to their lack of immediate, short term or long term physiological damage. My suggestion is that there should indeed be research but to provide meaningful, objective assessments of the existing masses of evidence. In around 1960 immunologist/developmental biologist Sir Peter Brian Medawar, OM CBE FRS, (my Dad's tutor, in fact!) stated that we had all the evidence to cure cancer – we just had to sort out what we'd got and link it all together properly. I'm sure he was right and the same is true here.

Of course the mechanisms of immunogenesis are of deep, deep interest – individually, custom tailored DNA sequences and so much more cry out for further research. Ideally we must find out how we can maintain the natural sensitivities at optimal capacity. It's no accident that illness often follows periods of great stress or intensive hard work.

Communal living in big groups was clearly a serious issue in Victorian slums. Gabriel Garcia Marquez' "Love in the Time of Cholera" depicts the same in South America. Cholera hit during economic recession, when resistance was low. But modern living conditions are so much improved and both diet and cleanliness are vastly better. (As I said before, this has allowed other issues like diabetes and obesity. My, we are stupid!)

In fact I said that "We cannot now waste time in creating in vitro laboratory models to describe these complicated interactions, which can have at best marginal relevance to this globally widespread problem." So we have models for optimal function and then for compromised. We demonstrate differences. We postulate mechanisms that could create the compromise.

This is good research, but it will not inform the solution to a problem resulting from a process invented by a 19th century charlatan (Jenner) and assumed since his time to be the fundamental prerequisite to enable the natural, four billion year developed immune system to function.

"We need to learn our vaccine science better". Yes and these days you can take a degree in Vaccinology, I know. And "figure out what are the risks versus the benefits". With methodologies such as used to test the MMR – by showing it had no worse an effect than other vaccines – there's no chance this will be done. The MIC call the shots, find the money and tie it to the results which support their industry. It is naive to think otherwise.

We need our immune systems to be keen and alert for any problems and not compromised by saturating them with a small group of antigens isolated from a small group of sometimes pathogenic organisms often many years ago and kept in culture, evolving, ever since. The science we should study is immunogenesis and not vaccinology as this latter assumes only one mode to develop one's immune defences. And that mode is fatally flawed.

The Scientist

I don't believe in maintaining a status quo simply to avoid rocking the boat. Hopefully my post is some evidence of that. However, I also don't believe in precipitous judgments when I can help it. I know you have read a LOT more on this topic than I have and I can only imagine what you've seen and what you've read have convinced you of your position. But so have the things I have seen and read in my lifetime. I am a cautious person. I don't believe that vaccines should be done away with and see them as a greatly beneficial aspect of modern medicine. BUT I'm not so blind as to assume they are harmless under every circumstance. And those circumstances in which they may not be, I want those studied and well understood. This kind of understanding would help us apply them in moderation to avoid unwanted effects while maintaining their benefits, and at the same time aid in development of future vaccines (which are an inevitability, so they may as well be designed well and safely).

I can see that my lack of full-fledged agreement with you is frustrating. I'm genuinely sorry I can't give you the support you're seeking beyond what I've offered. But at the same time, I'm equally frustrated you sound like you're missing my point. Not a criticism, just an occasion in which two people such as ourselves are talking "at" each other more than "with". Do I disagree with your approach to solve this issue? Yes– although not quite as much as opponents of further investigation would be. But then you disagree with mine also, so I suppose we're even and no need to get emotional nor take it personally.

Me

I don't look for agreement, I look for openness and am impressed to find it when I do. Faith based science and faith based medicine I grew up thinking were long since gone. How naive!

When I see a flawed argument I challenge it and provide counter evidence. In replying to your writing here I have sought to highlight illogical assumptions and their resultant derivations and to place logical, evidence based alternatives. I have seen none of them countered at all.

It's not personal – heaven forbid! But it is crucial for reasons of the Russian Roulette I outlined earlier (and that's probably a good measure of the odds of damage here) but also because it has the mindset of removing responsibility for health maintenance from the individual. It's all so archaic and voodoo. Not scientific at all!

The Scientist

I feel like I'm generally open and I'm waiting for science-based investigation to guide my opinions. I respect your opinion considering such practices medieval but I definitely do not. For me, they lie at the heart of good analysis and reliable data. So I guess we're just going to have to agree to disagree.

Me

I know you keep an open mind – "Medieval" is for your new chat about olde days illness (another blog post she'd made, elsewhere).

I do not call open, objective science medieval. For heaven's sake, it's what I do, so I could hardly slag it at the same time!

Archaic and voodoo is descriptor for vaccine medicine – a faith based malpractice.

Wow then even that didn't see the end of the discussion. She came back with a letter containing so many points and questions that I answered direct into her screed and sent it by email. Only then did I realise she was not English but based in Kentucky, which put her in a different context! Anyway:

Answers for the Scientist

Question – I'm just curious what you would specifically proffer in lieu of vaccinations though. If you bring up the topic of antibiotics, I'm right there with you and think that better alternatives are available, such as further investigations into utilizing antagonistic organisms to fight off other infections (e.g., combinations of yeast and lactobacilli to combat clostridium, etc.).

Answer – we already got there: good housing, diet, so nutrition, and clean water supply, life style, cleanliness, excellent emergency health care,

And while I can certainly fathom that vaccinations aren't perfect
No vaccine has yet been proven to be either effective or safe and it's a real balancing act trying to make an effective inoculation meanwhile not harming the patient, they have still prevented many illnesses

Nope – just look at the pre-vaccine decline in fatalities in first half of 20th century quite of few of which are potentially deadly. While I'm certainly a believer in not intervening with the job of the immune system unnecessarily, if the reaction is severe enough the immune system does kill people

Anaphylaxis you mean? Set off as potentiation to prior jabs and some sort of preventative or treatment is ideal.

Adrenaline shots

It's wonderful that you had mumps and didn't have any apparent lasting effects, but not every person is as lucky.

Luck didn't come into it – just normal good management – diet, warm cosy clean home

Same thing with a disease like rubella: there's plenty of people who are asymptomatic, which is wonderful for them. But what about the infants born with CRS?

Congenital rubella syndrome? If the mums had had normal infection as kids their lifelong immunity would have meant no possibility of catching it whilst pregnant and they'd probably have had passive immunity to pass to their kids. Use of vaccines is creating more and more problems, you see.

Or the ones who aren't born at all? People cry and scream over autism, but there are many more diseases than that.

Autism is not a disease as measles – it is a physiological state, akin to diabetes and still very hard to reverse to the prior non-autistic state.

And on that topic, while I can also fathom vaccinations playing a role in some individuals' aetiology, there are a HUGE number of environmental effectors that, like another commenter had suggested, probably each play minor cumulative roles.

I dealt with this point before – the vaccine is the initiator of the problem, the spanner in the works, other factors are ameliorators or even accelerators

I don't look for a Holy Grail of autism but look for an entire web of effectors, each web varying by individual.

A spanner is a spanner is a spanner.

Truthfully, I don't like all-or-nothing solutions.

"Keep spanner out of works" is not a solution – it's avoiding the problem in the first place

Vaccinations used in moderation, yes. Vaccines done away with entirely, no. Why? Because I'm not personally familiar with a better alternative.

See above

If you can give me one, aside from just letting people get deathly sick, do please. I'm very eager to learn. While I've seen some of the usefulness of antibiotics fighting off infection,

As the Chief Medical Officer noted today our overuse of antibiotics is running headlong into obviating their effectiveness. It's a long time since I first wrote decrying their prophylactic use in animal husbandry – chicken, pigs, cattle – and subsequent transfer of resistance plasmids to hospital situations.

they've been horrifically overused but the wonderful thing is that alternatives are already being developed, utilizing the natural antagonism certain microbiota have with one another as the bacteriocidal means. But viruses? Not that I'm aware of. Letting the immune system just battle it out is not a better solution in my mind.

Not "Just battle it out". We are so far better than that and, as I keep on repeating, the immune system is very, very sophisticated and well developed. I'd visualise developing generic immune system toning and reinforcement as opposed to the current system of hijacking and dedicating too great a part of the system in particular and artificial function.

For certain non-threatening illnesses, sure.

Illness is life threatening because of the sufferer's physiology and not because of a particularly evil microbe. Mind you, we are doing our best to construct such in our hospitals "although hospital infections from bugs such as MRSA and C.difficile have fallen, they are being replaced by other bacteria such as E. coli and klebsiella" as the "i" newspaper say today.

But for illnesses which have high morbidity and mortality, no way.

Not much else left, though, is there?

Nature isn't perfect, the immune system isn't perfect

It isn't? and death is extremely common.

Happens to us all

Heritability makes up for it in numbers.

Not if we're building up an enormous population of several generations weakened and compromised immune status, loss of materno-foetal transfer, loss of day to day adaptability and saturated with auto immune and allergic type problems. That's where we're heading. Most of us but luckily not everyone!

And then a repeat, as she sent back another sortofalistysortofaquestionysortofaposting. I obliged:

The scientist:

I've read through the pdf and I will think about it for awhile. I really would like to see research backing up claims. Hopefully you won't think too harshly of me, being a silly scientist

Now, now – covered that one, too. I'm a scientist and so understand evidence collection and, indeed, viewing. In fact I expect it.

who revolves around that kind of stuff. From my work and my understanding of autism, it is not a group of syndromes caused postnatally

There were no autistics before the 1940s. Where've they come from?

i.e., the time of most vaccinations excepting those given during pregnancy.

Vaccines to expectant mums is a very recent and deeply stupid development.

I'm a developmental biologist and I'm fairly familiar with how the central nervous system develops. There is a LOT of evidence which supports early effects in corticogenesis.

Like studies of foetal tissue??

This occurs within the first trimester. Now, this might not be all cases, but it's a sizable portion. Specifically, there appears to be an increase in neocortical proliferation targeting the radial glia within the ventricular zone underlying the cortical plate. This zone is gone by the neonatal period. So if there are effectors targeting proliferation of these cells, it must be very early indeed.

Please don't do the "Let's blame the mums" thing for this. It is so cruel.

I see vaccines, should they prove to have some effect on etiology in cases of autism, as one player amongst many. Perhaps triggering an autoimmune reaction

Certainly vaccination establishes autoimmune capacity as a major example of its collateral damage to the recipient's physiology.

which subsequently alters brain connectivity, triggering regression and more obvious signs of autism. Ultimately, autism = connectivity, though this is still only meagerly defined.

Or lack of connectivity!

But even if those individuals who regress following jabs show no signs of neocortical overproliferation, they are a minority.

It's a whole body syndrome – not just the brain. Analogous again to diabetes but with severe and irreversible neurological impacts. And a significant portion of autistics share a prenatal etiology,

Blame the mother's diet? Hey, then again there are those jabs she's getting nowadays. They must surely impact. But cross placental transfer? Dunno about that. Would have to look into it. Do you thing that Merck or GlaxoSmithKline might cough up the funding for such a study – and let you publish the results?

which I suspect is further complicated by other environmental effectors, like immune challenges

We're built to react to these – it's called the immune system.

steroidal agents, etc. There is a lot in this industrialized world which wasn't here several hundred years ago, agents which could prove powerful mutagens and teratogens.

From coffee to modern communication technology, sure, but when you examine the history the overwhelmingly clear initiator of autism and a range of other chronic syndromes is the archaic and never tested voodoo of vaccines.

And I think many such agents can fill those shoes (...she had ended).

And now so can I!

PS – There have to be footnotes:

1. *An analogy for the MMR epidemiology surveys: If you poisoned someone, say with Arsenic but were not sure whether the chemical caused the damage a comparison with a population fed on strychnine would tell that "Arsenic causes no increase in those suffering poison induced death". This would be true but would not show up how both chemicals were causing death.*

10. *Other vaccine induced malreactions fill a very long list – cot death, allergies, asthma, childhood leukaemia, eczema, ADHD, SSPE are a few of the more obvious.*

11. *Must go through whole piece adding notes where appropriate.*

12. *Yes, this is still a raw screed and requires in depth footnotes and references. I think though that their absence at first does not detract from its relevance.*

10 - John Humphreys and a little Tees

June 3, 2013

John the Turncoat, presenter, Today, "the UK's leading news/current affairs show", on BBC Radio 4.

Bless him, he seems a mild fellow, gently humorous, able to see the ironies and contrasts in life. Obviously grew up in the era when seeking knowledge and information and spreading it to the people was seen as a virtuous and glorious pursuit – surely to be rewarded in heaven. Well the weekly visit to chapel would have emphasised that benefit, for sure.

As a reporter he could do this – ferreting out scandals and misdemeanours for his local rag and climbing the journalistic ladder to BBC reportage where the same rules held true. He found he could lambast live these offenders and make them squirm in front of the whole nation.

So his role now as a latter day Joseph Goebbels must be quite galling to him. He's there as senior citizen of news and current affairs and as a figure of refinement and constancy. He has worked his way through the Andrew Gilligan and the Saville disturbances, and steadied the ship. But his insight has not just vanished – it's been banished. His questions are not just scripted – they're censored and the end result is like listening to a lobotomised version of his former self. I'm not saying, of course, that that's not what happened – "Could you come with us John, there's a nice consultant we'd like you to see about your condition", but I like to think that the old guy is still there, somewhere, behind this current projection.

But his present utterances are so often simply spouting, again and again, the propaganda of the centralist self perpetuating dogmatists.

Corporatist leaders, bankers and war-mongerers, like the inane and quite probably insane, muppet and puppet WilliamHague. These days sadly so often just voicing the opinions of his boss, Rupert Murdoch, master string puller that he is. Whatdoyoumean "He does not work for the BBC" – they work for him – have you not noticed yet?

So today, as I washed last night's crockery in the kitchen and brewed my aromatic Columbian Celestial Coffee, hand picked in the moonlight by a modern Inca workers' collective union and lovingly matured, dried and roasted by all knowing craftsfolk, I had to listen to John the Turncoat receive information from a smooth talking Teesider, telling us how he'd found reason to manage injection of MMR toxic cocktails into several thousand local teenagers because around 300 cases of "confirmed measles" had occurred in his patch.

Did John mention vaccine damage? Did John worry aloud about allergies, asthma, subacute schlerotising paraencephalitis (SSPE)? Did he mention autism? Did he question whether there is ever any measured benefit from the process? Did John suggest that the Swansea "epidemic" of measles had been a charade staged for the media to needlessly worry about mild childhood illnesses and rush their loved ones to the local clinics? Did John question the unexplained deaths of a 25 and a 17 year old, both in Swansea, quite possibly directly after their receiving the MMR jab resultant from that managed panic?

Did he mention those? Of course not.

"Pretty Polly. Pretty Polly."

11 - On infant dosed antibiotics and childhood obesity – setting the scene for the adult obesity plague?

22 November 2013

Margaret

More and more and more manure on antibiotics. Check out these maps. Of course, we know that the more doses of antibiotics given to children, the higher their BMI as they age! Shout it out: Use of antibiotics is CAUSING obesity, not vice versa.

"When we mashed up the data behind these maps, we confirmed the strong correlation between obesity and antibiotic prescription rates (we got an r of 0.74, for the statistically inclined). We also found a correlation between the states' median household incomes and antibiotic prescription rates: States with below-average median incomes tend to have higher antibiotic prescription rates. This makes sense, considering that high obesity rates correlate with low income levels. (You can see the data sets for antibiotic prescription rate, obesity, and median household income level here.)

Hicks and her team can't yet explain the connection between obesity and high rates of antibiotic prescription. 'There might be reasons that more obese people need antibiotics,' she says. 'But it also could be that antibiotic use is leading to obesity.'"

The states with some of the highest usage of antibiotics and rates of obesity: Well, of course, W, VA and MS — also the states with the most heavily vaccinated population of kids. Err, ya think the poor still are being used to make drug cos. rich in the U.S.? Ya think?

http://www.motherjones.com/environment/2013/11/mapsantibiotics-prescriptions-obesity-states

Margaret
Looks like this news will hit hard in the U.S "Antibiotics not for running noses, warn doctors". [Most] doctors won't like it!

http://www.phpnepal.org/index.php?listId=552&goback=.gde_1620737_member_58080634197099356619#%21

Chris Hemmings
Which is why "this issue" is taking a long time to highlight – there's clearly several factors acting to compound each of the inputting problems. Whilst the resultant damage is clear to everyone, pinning that on the "individual cause" is often well nigh impossible – there being NO SINGLE CAUSE.

So then it must be – what is/are the underlying problem(s) and what are simply compounders?

To me, antibiotics fall into the latter category such that if they are put aside the underlying problem(s) will remain present.

1 – Why are the antibiotics prescribed? IE – for what ailment?

2 – Why did the individual succumb to it?

3 – What is the vaccine status of each obese child?

4 – Is there a correlation between an individual's usage of antibiotics and obesity? Population studies do not illuminate this at all.

Yeah, antibiotics have for many decades be used in chicken feed so's they gain weight quickly as well as so they can be packed at high density. But this is prophylactic use which is less common for humans (altho' not wholly absent, I know!)

Margaret

Chris, so far, the epidemiology bears out the fact that antibiotic administration to children leads to higher BMIs as they age, especially when antibiotics are given to children under the age of six months. I have the links to the original article on this research published last year in Nature as well as to NIH's acceptance of the theories which underlie this conclusion. The biological explanation as to how antibiotics, in fact, do lead to obesity in children as they age strikes me as very, very sound.

I fortunately was born about a decade after antibiotics hit the market. By that time, doctors had discovered the hard way that these drugs actually may kill children. My brother almost died after a shot of penicillin given to him as an infant (1948 or 1949), and I myself developed a form of cancer (which I miraculously beat) from an antibiotic given to me in the early 1960s.

It became standard good medical practice back then NOT to give antibiotics to children unless their lives were at immediate risk of loss if antibiotics were not attempted. All of this had changed a bit by the late 1970s when tetracycline was discovered to take out teenage acne easily and, of course, in the late 1980s when vaccine mania began (since all vaccinations contain multiple doses of different antibiotics).

Antibiotics may do a whole lot more than just enable a body to get well more quickly from an infection, so I tend to favor judicious use of these drugs. But for Ceclor in the late 1980s, I would have had to undergo some extensive surgery on my sinuses. This was surgery that would have disrupted my life beyond description at that time (I was in law school) and which had a very low success rate anyway. Another point in that second article is the mention that "green" stuff which drains from the nose does NOT indicate an Rx for an antibiotic. From all that I know, whenever anything drains from the nose that is putrid green in color, doctors whip out that prescription pad and order up antibiotics as fast as they may write!

Chris

Yeah, my interest in antibiotics began with doing research on antibiotic resistance genetics within bacteria and its ease of cross species transfer (in "plasmids", being tiny, independent genetic units in the bacterial cell).

Conclusion was my advice to minimise antibiotic use and ban all prophylactic applications!

But (to paraphrase what I guess is the suggested rationale in relation to the results you quote) "the microbial biome in the gut is put out of balance and takes time to reassemble" does not really EXPLAIN why obesity follows. Truth is that every meal alters the gut microbial content – there is a constant adaptation to ITS environment (Sugars, carbohydrates, other bacteria, green vegetables.....)

So although the effect of an obesogenic diet will almost certainly be compounded by an already skewed gut flora it will still generate obesity when antibiotics have not been administered. Wheat flour is highly obesogenic, for example, as it has practically the highest Glycemic Index (GI) of any food – higher than table sugar. Eat wheat and your blood sugar shoots up **and** remains high until your next burger, cookie or organic wholewheat sandwich.

High blood sugar leads to obesity and diabetes…….

Margaret

"These effects are broad across vertebrate species, including mammals (cattle, swine, sheep) and birds (chickens, turkeys), and follow oral administration of the agents, either in feed or water, indicating that the microbiota of the gastrointestinal (GI) tract is a major target. That the effects are observed with many different classes of antibacterial agents (including macrolides, tetracyclines, penicillins and ionophores) indicates that the activity is not an agentspecific side effect, nor have the effects been observed with antifungals or antivirals.

The vertebrate GI tract contains an exceptionally complex and dense microbial environment, with bacterial constituents that affect the immune responses of populations of reactive host cells[8] and stimulate a rich matrix of effecter mechanisms involved in innate and adaptive immune responses[9]. The GI tract also is a locus of hormone production, including those involved in energy homeostasis (such as insulin, glucagon, leptin and ghrelin) and growth (for example, glucose-dependent insulinotropic polypeptide (GIP) and glucagonlike peptide 1 (GLP-1))[10]. Alterations in the populations of the GI microbiota may change the intra-community metabolic interactions[11], modify caloric intake by using carbohydrates such as cellulose that are otherwise indigestible by the host[12], and globally affect host metabolic, hormonal and immune homeostasis[13]. Full (therapeutic) dose antibiotic treatments alter both the composition of the gastrointestinal microbiota[14] and host responses to specific microbial signals[15]. In combination with dietary changes, antibiotic administration has been associated with changes in the population structure of the microbiome."

http://www.ncbi.nlm.nih.gov/pmc/articles/PMC3553221/

"In related work, Dr. Leonardo Trasande, Blaser and colleagues analyzed data from over 11,000 children in a British study, examining body mass and antibiotic exposures during infancy. The results appeared on August 21, 2012, in the online edition of the International Journal of Obesity.

"The researchers found that children who were given antibiotics during the first 6 months of life were more likely to have a higher body mass and were more likely to be overweight by 3 years of age than those who weren't given the drugs. Exposure between 6 and 14 months wasn't associated with body mass index at any time point. While exposure to antibiotics between 15 and 23 months appeared to affect body mass index at 7 years of age, none of the exposures were linked to being overweight or obese at 7."

http://www.nih.gov/res.../october2012/10012012antibiotic.htm

Chris
OK, I read that, Margaret, and can only see support of my observations – at 7 years old, the impact of earlier antibiotic use falls off, it seems, and has no impact on predicting, ie precipitating, future obesity.
Use in poultry is, of course, continuous from chick to dispatch, being in their food.

The complexity of gut biome is why I use the phrase "Bio-medical Ecology" to cover these interactions, in the human system. I've tried to push this for years and yet, a few months back, it seemed as if modern science was only just realising the gut contained a bacterial flora at all. What woke them up? Why the chance of producing some new vaccines…….

http://bmeandothersciences.wordpress.com/2013/06/07/so-howdo-we-contract-typhoid-enteric-bacterial-sagas-and-those-cash-richvaccinators/

Margaret

Chris: Thank you so much for that article! But for the money infused into it, of course, I always wonder why research which focuses on how to "boost" human immune response to disease must concentrate its efforts on vaccination-created changes to the human immune system. No paradigm could be more outdated or dangerous.

Here is the full text to that article on the 11.5K kids examined from the International Journal of Obesity (London). You might find reading ALL the results to be more telling than just that conclusion which NIH reports above.

"This longitudinal study found that early-life antibiotic exposure was associated with subsequent increases in body mass. Of the three time windows analyzed, only exposure during the period before 6 months of age was consistently associated with increases in body mass. At 38 months, children who had been exposed to antibiotics during this earliest period had significantly higher standardized BMI scores, and were 22% more likely to be overweight than children who had not been exposed.

"In contrast, exposures after 6 months were not consistently associated with body mass increases. Those in the 6- to 14-month period showed no association, while those in the period 15–23 months were significantly associated only with elevated standardized BMI score at 7 years, but not with consistently elevated scores in the interim.

"Our finding of an association of antibiotic exposure at <6 months with later-life body mass is consistent with a prior report. It adds important evidence that exposure timing matters. We also add a test of spuriousness, with the finding that exposure to non-antibiotic medications during the same early windows is not associated with elevations in body mass. This makes it less plausible that exposure to antibiotics reflects a greater receptivity to medications, which is also correlated with increases in body mass. Unlike the previous study, however, we did not find an association with antibiotic exposure at <6 months persisting to 7 years of age.

"Perhaps this reflects differences in the antibiotics used in the two samples, or different doses. Given that intravenous antibiotics are used in these first 6 months of life (often for neonatal sepsis), antibiotic type (that is, Gram-positive or Gram-negative/anaerobic coverage) and route of administration (intravenous or orally administered antibiotics) might have differential effects on gut microbiota composition and development. This is consistent with a recent analysis finding associations of intravenous vancomycin, but not amoxicillin, treatment in adults with the development of obesity. Alternatively, our failure to find a significant association may simply reflect our somewhat smaller sample size."
http://www.ncbi.nlm.nih.gov/pmc/articles/PMC3798029/

Chris
Mm, yes, and you could have put up this, their conclusions, as well:

1. "Exposure to antibiotics during the first 6 months of life is associated with consistent increases in body mass from 10 to 38 months.

- "Exposures later in infancy (6–14 months, 15–23 months) are not consistently associated with increased body mass.

- "Although effects of early exposures are modest at the individual level, they could have substantial consequences for population health.

- "Given the prevalence of antibiotic exposures in infants, and in light of the growing concerns about childhood obesity, further studies are needed to isolate effects and define life-course implications for body mass and cardiovascular risks." And then:

"While the composition of the microbiota of adults appears relatively stable, the microbiota of children may be considerably more variable and more vulnerable to antibiotic perturbation.

"Increased risks from childhood antibiotic exposure for atopic dermatitis,asthma and inflammatory bowel disease have been reported."

Whilst in Denmark they found:

"Antibiotics during the first 6 months of life led to increased risk of overweight among children of normal weight mothers and a decreased risk of overweight among children of overweight or obese mothers ." At age 7.
http://www.ncbi.nlm.nih.gov/pubmed/21386800
There's clearly stuff happening and also it's abundantly clear that it has not been looked at anything like well enough yet. I'm always aware of the fact that the whole alimentary canal is external to the

human body, just contained within it, and so the whole gut lining is body surface tissue, akin to the skin, interacting with the environment surrounding it.

I think antibiotics might have saved my life once (infected arm injury) as well, and so I am also ambivalent about them. My early research conclusions I still maintain – but sadly far too much prophylactic use continues and is compounded by its extension into prolonged use in human medication. This can only decrease their effectiveness in emergency situations, select for even more powerful antibioticresistance plasmids and provide an environment where problems like obesity may well be encouraged or compounded.

Margaret

I, too, have much for which to thank antibiotics.

On the other hand, PROPHYLACTIC administration of chloramphenicol to me via twice daily injections, when I was about a six year old, caused me to develop a form of cancer, aplastic anemia. I am one of the extremely rare kids with this drug-induced disease from the early 1960s who has lived to tell the story. Interestingly, there were NO medicines other than iron and some "sugar formulations" available back then, so one might say that my own body won that battle.

What precipitated the doctor's decision to use this drug? His fear that the SOS, recently forced onto me then, was about to cause some "awful" disease in me. Two of my friends at that time had developed glomerulonephritis from the SOS; both almost died and had to spend almost a year in bed while recovering.

One persistent problem with drugs: They always are used to make money and the careers of scientists first. If only mankind were able to cut out the greed and use drugs only when they are indicated and known to be of some genuine benefit to human health.

Chris
Pardon my dumbness but "SOS"?

Margaret
Sabin Oral Solution, aka OPV today.

Want to be made even sicker: The three doses (the little pink sugar cubes) were handed out at local high schools (or other public buildings) after church on three consecutive Sundays. SOS then stood for "Sabin Oral Sunday." Man, those memories almost bring me to vomit. Yet, that type of hoodwinking of the pubic by health authorities goes on in spades to this day.

My mother was absolutely frantic when she discovered that she could not get me exempted from the SOS. Most people in the know back then, I think, were well aware of the fact that this vaccination was a "death shot." My family moved to NYC a couple of years after the SOS campaign. My doctor there was Harvard educated, not a pediatrician but a cardiologist (our family doctor). He told my parents that he managed to get each and every pediatric patient in his practice exempted from the SOS. He called Salk, Sabin, and the entire polio vaccination fiasco a "sad joke."

For the sake of history, this doctor also had a unique way of having his pediatric patients vaccinated with the smallpox vaccination (no one got out of that one back then): He chose a spot on the back of the forearm where the scar never would be seen. How kind of him.

Chris

Name changing, shape shifting, alchemical practices. It's all so Medieval but, hey, the irony of "SOS" is not lost on me! Help, help.

12 – The Jab Patrol? Tightening the Grip?

December 15th, 2013.

Claire asked us a question:

"Hello everybody just wanted to ask if anyone knows the law on having children vaccinated in the UK? I am currently expecting and I want to know my rights for my and my child's health. Worried I could be taken to court if I do not get the MMR vaccine etc!!!"

Vidhya

From what I recall you can easily opt out and still attend school. No need to prove religious grounds or anything like that. Don't know more details as I moved to the US.

Chris Hemmings

Yeah, they only use social, medical and misinformation pressure. No legal governance except, oddly, in the case of the parents separating. Recently an estranged father obtained court backing to force MMR on the two teenage daughters who lived with the mum. Prior to separation he had supported non-vaccination!

Claire

Yeah Chris that's exactly why I asked I followed how that case went and it concerned me that the court can make a ruling. I guess it was due to the fathers case but this was exactly why I got concerned is this the way things are moving due to big outbreaks which have occurred.....

Chris Hemmings

Yes, it's an alarming area. The undercover, covert legal system, termed "family courts" but whose actions are behind closed doors and no press or individual permission to carry information beyond the hearing.

If "the father's case" was "good" then it upturns so many legal precedents in this country. I believe both the mother and the two girls intend to go on fighting the judgement. I hope so because if it becomes a new accepted precedent then it's one tiny step from state mandation – one could envisage "The Jab Squad", syringes primed, cruising the streets looking for any remaining undamaged kids.

Not trying to worry you......

Claire

 Problem is when registering a birth we automatically become subject and a vessel of the state!! The state owns our ass! Contemplated not registering the birth to avoid this however this would cause wide ramifications for adulthood, having a bank acct, national health/ insurance, a passport and so fourth!!! Sad facts are like you said this may be paving the way for the future control therefore have got you either way!!

Chris Hemmings

My morality says change the system from within, but my instincts these days say the opposite. I've followed the former but corporate and institutional pressures are enormous and contain no grain of humanity to ease them. They bear no compromise that I can see. The stakes are so high.

13 - Interesting and scandalous about "the swine flu pandemic" and new Nordic mythologies coming from their research workers.

22nd November, 2013

Yes, an interesting chat, indeed, being an exemplar in the wheeler dealing, the cavalier behaviours adopted by industry operators and the lack of objective controls to such actions or subsequent come back upon their personal or corporate heads:

Sandy

Interesting and scandalous about the "swine flu pandemic".
http://www.bmj.com/content/340/bmj.c225.full:

"H1N1: now entering the recrimination phase"
"If influenza was a rock band how would it rate its latest release, H1N1? Not too well, I suspect, despite the greatest pre-publicity since—well, its previous release. And it all started so promisingly, in Mexico, whose population had been decimated by the very first outbreak of Spanish flu (and smallpox and measles), courtesy of Cortés and his conquistadores.

"The new lineup—two parts pig, one part human, and one part bird (The Chimerical Brothers?)—looked brilliant on paper. Once the international tour began, all eyes were on the southern hemisphere for pointers as to how things might play out in the northern hemisphere winter. So what happened next?

"For England, many more misses than hits. Since last August, the consultation rates for flu-like illness have hardly budged above the baseline threshold (doi:10.1136/bmj.c170). They're now less than half that rate and falling. Even the most generous assessment couldn't attribute this happy state of affairs to either the use of oseltamivir (Tamiflu) or vaccination against swine flu. Both interventions are now uncomfortably under the spotlight.

"This week we publish the latest in a series of letters looking at the downsides of distant diagnosis by algorithm. Catherine Houlihan and colleagues from Newcastle upon Tyne reviewed eight cases of potentially life threatening conditions where diagnosis and management were delayed because of an initial incorrect diagnosis of swine flu (doi:10.1136/bmj.c137). Last August we published a similar series from Middlesbrough (BMJ 2009;339:b3365, doi:10.1136/bmj.b3365). Once this pandemic is over, it would be interesting to tot up the national total of clinically significant diagnoses that were initially missed because of the too ready diagnosis of swine flu. Meanwhile, European governments, including the UK's, are trying to offload their surplus stocks of swine flu vaccine as vaccination programmes are canned.

"The search for scapegoats has already begun. The chairman of the health subcommittee of the Council of Europe's parliamentary assembly has called for an investigation into the role of pharmaceutical companies in the current pandemic (doi:10.1136/bmj.c198). His charge: "To protect their patented drugs and vaccines against flu, pharmaceutical companies have influenced scientists and official agencies, responsible for public health standards to alarm governments." Meanwhile, the revelation of undeclared competing interests of Professor Juhani Eskola, an adviser to WHO's Strategic Advisory Group of Experts (SAGE), has come as a gift to conspiracy theorists. SAGE advises member states on vaccines; GlaxoSmithKline, manufacturer of Pandemrix, is the main source of income of Professor Eskola's employer (doi:10.1136/bmj.c201).

"Recriminations of a different kind in Liverpool. In "The Price of Silence," Jonathan Gornall's article on the Liverpool Women's NHS Foundation Trust, he claimed that 12 compromise agreements entered into by doctors there contained gagging clauses (BMJ 2009;339:b3202). The trust's chairman replied that such agreements affected only two doctors (doi:10.1136/bmj.c144).

"But Andrew Bousfield, whose father had been banned by the trust from going public with concerns about management and patient safety, had specifically asked for information relating to doctors, and under direction from the Information Commissioner the trust provided him with 12 redacted copies of compromise agreements (doi:10.1136/bmj.c145). So who's right? We need an adjudicator to check the unredacted forms."

Jyrki

The BMJ slightly messed up the info about the Finnish THL's [Finnish National Institute for Health and Welfare] GSK [Glaxo Smith Kline] money source. We got the documents via the Finnish Freedom of Information act. Confusion comes from the fact that Danish media was the first to publish the financial ties – most of media in Finland for some reason is very pro THL and pro vaccine.

Also we later found out the contract is not €6.2 million but more than
€10 million. And the conflict of interest continues – Dr. Terhi Kilpi, leader of GSK-funded trial, is now sitting at the WHO vaccine safety committee, giving statements on the safety of GSK's vaccines.

Now THL has bought HPV vaccines from GSK and the HPV vaccination campaign has started in the last weeks. The misinformation fed to the media by THL and the ministry of health is even worse than it was at the time of Pandemrix, and the media is even more uncritical of THL's position than at the time of Pandemrix.

"Documents acquired through the Danish Freedom of Information Act by the Danish daily newspaper Information show that Juhani Eskola, a Finnish vaccines adviser on the WHO board, has received £5.6m (€6.2m; $9m) for his research centre, the Finnish National Institute for Health and Welfare. The money, from GlaxoSmithKline for research on vaccines during 2009, is the institute's main source of income." http://www.bmj.com/content/340/bmj.c201

During the times of Finlandization (1), there was a media phenomenon when "the Soviet Union mustn't be criticized in Finland" due to foreign political reasons. Sometimes the consequence of this was that other Nordic (e.g. Swedish) media was reporting things which "could not" be said by Finnish media, and then Finnish media could quote these things as the "cat was out of the bag", so to speak. Apparently vaccines are even more of a sensitive issue than foreign policy – Swedish media quoted the Danish report, but I think only one Finnish newspaper did a small news piece on the Eskola story. Well then there was the public broadcasting corporation which uncritically gave Eskola a place to describe how he thinks things went down.

1. http://en.wikipedia.org/wiki/Finlandization

Mel
Wow, someone with some guts!

Jyrki

By the way, BMJ also published one of my comments on the topic of narcolepsy [as connected in Nordic areas with the H1N1 flu jab] in a printed issue. Don't remember exactly which one, maybe this: "Adjuvanted pandemic swine flu vaccine and narcolepsy" BMJ :

http://www.bmj.com/rapid-response/2011/11/03/adjuvantedpandemic-swine-flu-vaccine-and-narcolepsy

Sandy

Very interesting posts Jyrki and excellent comment in BMJ

Jyrki

Or maybe it was this one:

http://www.bmj.com/rapid-response/2011/11/02/whos-lackdisclosure-financial-ties

For a current update, sad to say, WHO's policies today appear to be even more closed and secretive than they were back then – [when] I was able to obtain Dr Eskola's confict of interest report and could see that he hadn't reported the GSK funding, but this year, repeated requests for Dr Kilpi's conflict of interest report have not produced any results except vague email responses from an unidentified individual claiming that the documents are not public.

Sandy

Yes there's less transparency and more censoring of info as time goes by. Great comment here Jyrki. It never ceases to amaze that however much interest conflicts/financial ties are exposed, the ones who are involved don't seem to care. Probably because they have political and pharma backing.

<u>Jyrki</u>

By the way, when looking into the HPV vaccine, I noticed the SPC had a lie or incorrect information about using a "placebo-controlled" trial – when I checked the trail on clinical trials, it was the one carried out in Finland where an experimental hepatitis A vaccine was used as control.

And here's the article which appeared in print in BMJ – actually my text was just a short quote from the latter of the two comments linked above:

"It is sad to see that even in April 2010, the World Health Organisation, WHO, has not yet changed the policies so as to be more open regarding the publication of financial ties of their WHO advisors".

http://www.crn2.org.br/pdf/artigos/artigos1277295220.pdf The other comment, posted on the article after my comment, is from Margaret Chan, DG of WHO!

<u>Chris</u> (I finally arrived to join in!)
Additionally, on Eskola from his WHO biography:

"During the years of 2000 to 2003, he worked as Senior Vice President at Aventis Pasteur in Lyon, France, being responsible for global medical affairs and clinical development of all vaccines in the company."

Interestingly/amusingly:

"During recent years his main interest has shifted into …… vaccine safety issues, and into modelling and measuring the impact of wide scale vaccinations."

Well, who'd have thought it??

http://www.who.int/immunization/sage/members/bio_eskola/en/index.html

Sandy
Jyrki they used false "placebos" for both Cervarix and Gardasil HPV vaccines: aluminium, hep a vaccine, you name it….

What beats me is that they often say "placebo" instead of "control", they surely know that this is grossly misleading.

http://us.gsk.com/products/assets/us_cervarix.pdf
http://www.merck.com/product/usa/pi_circulars/g/gardasil/gardasil_pi.pdf

[This use of language is, indeed, interesting. Placebo is non-functional substitute for a reagent, used to provide a non-altered base-line effect. Thus an alcohol free lager could be placebo in an examination of the impact of alcohol on behaviour of lager drinkers, whereas you could not use cider or vodka as, although different from the experimental reagent (lager) both contain alcohol. Sorry for such a pedantic descriptor, but, evidently some experimental scientists require this revision exercise.

In such an experiment, the alcohol free lager group is thus also a realistic control.]

Sandy continued:

A few Norwegian newspapers most surprisingly published (in 2009) some of my articles about Pandemrix. Here's one giving details about ingredients, formulation and administration:

"PANDEMRIX H1N1 VACCINE – AN UNWISE CHOICE?

Pandemrix has recently been suspected of increasing the risk of the auto-immune disorder narcolepsy. This vaccine has the lowest concentration of virus antigen of the H1N1 vaccines which are available on the market.

SQUALENE

Because of this the controversial adjuvant squalene (AS03) has been included in order to boost the immune response. (NB. Immune response is not the same as immunity!) Virus antigen is considerably more expensive than squalene.

Squalene has been suspected of increasing the risk of auto-immune disorders, including the Gulf War Syndrome, narcolepsy, multiple sclerosis and several others. Squalene adjuvants are routinely used to induce arthritis in rats in a laboratory setting.

It is known to cause more pain and inflammation at the injection site than squalene-free injections.

Because squalene is an oily substance the emulsifying agent polysorbate 80 (Tween 80) is added in order to blend it with the aqueous phase containing the virus antigen.

POLYSORBATE

Polysorbate has been suspected of increasing the risk of anaphylactic shock. Research shows that it causes reproductive disturbances in rats and infertility in mice. It is suspected of having carcinogenic and mutagenic properties.

Because it makes the blood-brain barrier more permeable, it is used in special injections to facilitate the passage of certain substances through the barrier into the brain tissue (drug targeting).

This property is obviously unwanted in vaccines. Polysorbate in Pandemrix may facilitate the passage of mercury in thimerosal and other substances through the barrier into the brain. Small children are especially vulnerable as they have underdeveloped blood-brain barriers. It is therefore no doubt unwise to include polysorbate 80 in injections for children and pregnant women.

Because of polysorbate's effect on the fate of mercury and other substances, the actual composition of vaccines should be considered in addition to the amount of mercury and other ingredients

EGG ALLERGY
The vaccine may not be given to people who are strongly allergic to eggs. This does not appear to have been considered when Pandemrix was chosen for mass vaccination.

SINGLE DOSE MORE EXPENSIVE THAN MULTI-DOSE VIALS
All injections may normally be manufactured in the form of single dose prefilled syringes or as multidose vials. Single dose products are more expensive to manufacture.

PRESERVATIVE: MERCURY

Single dose injections do not normally require preservatives, but these are necessary in multidose products in order to reduce microbial contamination after they have been opened. The preservative in Pandemrix is thimerosal which contains almost 50% mercury.

The presence of mercury is extremely controversial. It is a known neurotoxic that can weaken the immune system and cause neurological damage. Thimerosal has been shown to cause autismlike symptoms in laboratory animals. There are strong indications that mercury is the cause of autism in human beings.

UNEVEN DOSES
The vials of Pandemrix contain ten doses each of 0.5ml. Before withdrawal of each dose the vial should be thoroughly shaken.

Microbial contamination may occur in connection with the administrator's technique and from the surroundings, especially when there are many people present. This risk is less in the case of single dose injections. A higher degree of competence may therefore be required for administrating multidose injections.

The virus antigen is in the form of tiny, invisible particles dispersed in the vaccine. Irrespective of the administrator's technique regarding shaking, the correct amount of virus antigen in each 0.5ml dose cannot be guaranteed. (In the case of prefilled injections it is possible to regularly analyse and control the amount during online processes).

In the case of the multidose vials of Pandemrix, even if the doses of 0.5ml are accurately measured, one cannot know if one receives the correct amount or strength of virus antigen – or if it is the same as that which the next person in the vaccine queue receives!

TO SUMMARISE

If the authorities had chosen a more expensive single dose product it would not have been necessary to include mercury, there would be less risk of microbial contamination and a greater possibility that each dose has the correct strength of the virus antigen and other ingredients.

If the authorities had chosen a vaccine which had sufficient strength with respect to virus antigen, the inclusion of squalene and polysorbate 80 would have been unnecessary."

Jyrki I read somewhere Dr Eskola was part of a process of developing a new tuberculosis vaccine.

Also I think another thing he was involved [with] was some kind of financial model, where governments would cover the losses of vaccine manufacturers when a vaccine development process fails, but the companies would keep the profits when a marketing authorization is gained.

What the justification for doing those things on taxpayers' expense is beyond me. Though it's hard to know on whose expense the THL people do what they do, as they're so secretive about the money traffic while at the same time declaring to the world how important transparency is - as they wrote at their Lancet article: http://www.thelancet.com/journals/lancet/article/PIIS01406736(11)60690-9/fulltext

Chris

New TB vaccine reference is in the WHO biography I quoted, above.

Jyrki

For the Deputy Director General at THL, Dr Eskola [has been] very rarely seen in Finnish media after [the]Pandemrix narcolepsy [scandal came to light

According to media reports, HPV vaccine coverage based on girls' and parents' written notices will be around a third in some places up to two thirds in some other places. This despite (or maybe more probably because of) media being very one-sided on the properties of the HPV vaccine and the THL initiating a mass letter mailing campaign with pink letters [sent] to all girls and their parents. According to THL the coverage for other normal vaccine schedule vaccines traditionally has been > 95%, so looks like things may be changing.

Anyone who can say what HPV vaccine coverage is in other European countries or elsewhere?

[And the conversation lapsed. Presumably no-one had the information.

But, yeah, interesting AND scandalous!]

www.ingramcontent.com/pod-product-compliance
Lightning Source LLC
Chambersburg PA
CBHW080657190526
45169CB00006B/2158